THE BUSINESS CASE FOR RENEWABLE ENERGY

A Guide for Colleges and Universities

Andrea Putman

Michael Philips

Library of Congress Cataloging-in-Publication Data

Philips, Michael.
   The business case for renewable energy : a guide for colleges and universities / by Michael Philips and Andrea Putman.
      p. cm.
   ISBN-13: 978-1-56972-036-3
   1. College facilities--United States. 2. Renewable energy sources --United States.   I. Putman, Andrea, 1961-   . II. Title.
   LB3223.3.P486 2006
   378.1'960973--dc22

                                            2006010794

APPA
Alexandria, VA
www.appa.org

National Association of College and University Business Officers
Washington, DC
www.nacubo.org

Society for College and University Planning
Ann Arbor, MI
www.scup.org

Printed in the United States of America

contents

FIGURES

Climate change is one of the greatest emerging risks to the security, health, environment and sustainable economic prosperity of the nation and the world. Carbon dioxide levels in the atmosphere have been increasing rapidly to levels not seen in thousands of years, mostly as a result of greenhouse gas (GHG) emissions caused by human activity from fossil fuel consumption. These gasses warm the atmosphere, and more importantly, the oceans (where 90 percent of the sun's energy is absorbed), thus having a major impact on global climate patterns.

The overwhelming consensus of climate scientists is that climate change is already here and we are not doing enough to reduce and mitigate it. Reversing the impacts from human-induced GHG emissions to date may take a century (or more) and require GHG emissions reductions on the order of 70–80 percent from present levels, so we have to start mitigation and adaptation now, and attempt to reduce our GHG emissions without shocking our economies and threatening our financial stability. To accomplish this, we must rapidly reduce our dependence on the world's supply of fossil fuels to meet our energy needs. Such action would bring many benefits, including cleaner air, less volatility of fuel costs, and more energy security by reducing the need for energy imports.

The consequences of inaction are already with us and rapidly accelerating: climate disruption reflected in natural disasters (more intense floods, hurricanes, and droughts), sea level rise, shifting ocean currents, changing growing seasons and ecosystems—and the economic and political dislocations that follow. The 2005 "Climate Change Future: Health, Ecological, and Economic Dimensions" report by experts at the Harvard Medical School even links GHG-induced climate change to the recent rapid increase of asthma in American cities and suburbs and the spread of West Nile Virus, which is not good news for our already overburdened system of financing health care costs.

How will higher education in the United States be affected? There will definitely be winners and losers as a result of climate change, and the higher education system in the United States will not be immune to these impacts, risks, and opportunities. Climate change not only impacts the environment, but also our energy, economic systems, and financial systems, including potentially both public and private financing of higher education. Colleges and universities will face increased operating costs for energy, travel, property and casualty insurance, employee absenteeism, health care costs, decreased worker productivity due to increased illness or shortened life expectancy, and business interruption losses (look at the impact of just one intense natural disaster—Hurricane Katrina—on Tulane, Xavier, and Dillard universities).

There are plenty of examples of how the financial world is paying close attention to the risks of climate change. A major reinsurance company notes that the amount of insured losses from natural disasters has been doubling each decade for the past 40 years, and Allstate recently announced a plan to drop most homeowner insurance policies in coastal New York State due to climate risks. Recently, major U.S. financial firms such as Bank of America, Citigroup, Goldman Sachs, and JP Morgan Chase have announced climate change policies and/or have voluntarily agreed to reduce their GHG emissions over the coming years. Other firms, such as UBS, Citigroup, and HSBC, have purchased renewable energy, pledged to become "carbon-neutral" in their operations (HSBC has achieved that already in 2005), or established renewable energy investment funds.

A recent survey of the CEOs of the 500 largest companies in the world by a group of 155 institutional investors representing $21 trillion in assets indicated that 75 percent of CEOs felt climate change was a major business financial issue/risk, but that only 25 percent of these firms were doing anything about it. On the other hand, many investors are turning climate change risks into business opportunities. Several investment groups have recently taken an interest in climate change and carbon trading investment strategies, including a joint venture recently set up by a major U.S. insurance company and a large hedge fund focused on investment strategies and opportunities in the area of environment and climate change.

For business officers in higher education institutions, improved risk management of climate change will have an increasing priority. Likewise, managers of higher education endowments will have to start considering climate risks/opportunities embedded in their investments and operations. With rising energy costs, many institutions have started to harvest the energy efficiency "low hanging fruit" with lighting and equipment upgrades. However, our experience in New Jersey higher education indicates that further cost-effective measures could yield 15 percent–45 percent reductions in energy use. Our Energy Action Plans for 10 New Jersey campuses included more than $73 million in energy efficiency recommendations agreed upon by consultants and campus facilities managers. Most improvements, eligible for subsidies from the outstanding Clean Energy Program of the New Jersey Board of Public Utilities, will reduce operating costs and produce additional revenues from sales of renewable energy certificates.

Where does renewable energy fit in this picture? Renewable energy will be an energy source as long as Earth's ecosystems continue. Fossil fuels, our present major energy source, are limited in supply and extracting and using them is causing global warming as well as damaging the environment in many other ways. Renewable energy plays a central role in the necessary transition from a fossil fuel-based energy paradigm to a newly emerging renewable paradigm we are only beginning to perceive and understand.

Higher education is uniquely positioned to play a key role in this transition. As former New Jersey Governor Thomas Kean said when he was president of Drew University, well before he led the 9/11 commission that pointed out U.S. vulnerability because of dependence on foreign oil, "if our colleges and universities don't take the lead in moving us toward a sustainable world, then who will?" This led to all 56 New Jersey college presidents agreeing to voluntarily reduce campus greenhouse gas emissions to below 1990 levels by 2005. (Results of their efforts will be reported about the time this book is published.)

The leadership role of higher education is especially important because many colleges and universities have reported doubling and tripling of their energy rates and costs over the past decade, so investments in this area can be profitable as well as good for the planet. For example, Harvard University recently set aside a $5.5 million fund for energy efficiency investments in 10 on-campus projects, and the rate of return on this investment is 38 percent per year! Institutions are also interested in renewable energy sources such as solar, wind, and hydro because they have no fuel costs, thus allowing accurate estimates of long-term energy costs without the volatility of fossil fuel market forces beyond the institution's control. Examples of this approach of reducing greenhouse gas emissions by using renewable energy are growing in leaps and bounds in higher education, and many are documented in this book.

In our experience, there is great leadership potential for a paradigm shift regarding energy use in our higher education institutions. Presidents, trustees, and financial officers will back it because they realize the strategic and risk management value of renewable energy, as well as the financial benefits. Facilities directors will back it if they can see how it improves their energy efficiency, reduces operating costs, and leads to better buildings. Faculty will support it based on insights from their disciplines and across disciplines; students will support it when their teachers and mentors help them put together an encouraging picture of a future based on a different paradigm. Until now, there have been powerful books, distinguished champions, and dynamic movements to inform and inspire the academic community with the best case for a renewable paradigm, but there has never been a good explanation of the business case—"getting the numbers right," until this book.

Leaders in APPA, NACUBO, and SCUP caught this wave of change and commissioned this excellent, comprehensive guide by Andrea Putman and Michael Philips. The thrust of U.S. colleges and universities to change the existing energy paradigm, plus major basic and applied sustainability research, deserves to earn greater trust and support for higher education from the American people.

The other benefit of this leadership is that it fits the educational role of American higher education of leading by example. Much of the economic success of this nation has been built on the innovations and research that have come out of our excellent educational system. By practicing what we preach, and being leaders in adopting emerging renewable energy technologies, we can continue this proud tradition of innovation, help protect the long-term financial viability of our universities and colleges, and protect our environment for future generations of students, faculty, and administrators. Higher education administrators need to be made aware of the financial risks and opportunities presented by climate change in order to properly position their institutions for continued leadership on renewable energy, to serve their students, trustees, and stakeholders with correct fiduciary responsibility, and to maintain and protect the financial stability of higher education. Anything less is unsustainable in the long term.

John Cusack
Executive Director
New Jersey Higher Education
Partnership for Sustainability (NJHEPS)

Dr. Donald Wheeler
Immediate Past President
New Jersey Higher Education
Partnership for Sustainability (NJHEPS)

Many thanks to our sponsors which include New Jersey Higher Education Partnership for Sustainability, U.S. Department of Energy, International City/County Management Association, Chevron Energy Solutions, American Council on Renewable Energy, and National Wildlife Federation.

We sincerely thank the following individuals for their guidance and invaluable insights:
    Jeffrey Baker, U.S. Department of Energy
    John Cusack, New Jersey Higher Education Partnership for Sustainability
    Alden Hathaway, Environmental Resources Trust
    John Kunz, Environmental Resources Trust
    Henry Mauermeyer, New Jersey Institute of Technology
    Daniel Sze, U.S. Department of Energy
    Don Wheeler, New Jersey Higher Education Partnership for Sustainability

Special thanks to the following individuals for providing tours and detailed information on their institutions' renewable energy projects:
    Tom Brown, Kate Rodriguez, James Valiensi, Lynn Wiegers,
        and Joshua Gallo of California State University, Northridge
    Blair Doane, Los Angeles Community College District
    Ferman Milster, University of Iowa
    Frank Vitone, Pierce College
    Cheryl Wolfe-Cragin, Oberlin College

Thank you to the following individuals who generously provided us with information and advice:
    Anthony Amato, ERG
    Brent Beerly, Community Energy, Inc.
    Bonny Bentzin, Arizona State University
    Chip Bircher, Wisconsin Public Service Corporation
    Elliot Boardman, Peak Load Management Alliance
    Kirk Bond, Aspen Systems
    David Brixton, Arizona State University
    Claire Broido-Johnson, SunEdison Corporation
    Adam Browning, Vote Solar Initiative
    Tom Burke, Cerro Coso Community College
    Mark Byrd, Stevens Institute of Technology
    John Chase, American Wind Energy Association
    Matt Clouse, U.S. Environmental Protection Agency
    Gary Colella, SUNY Syracuse
    Chris Cooke, SunEdison Corporation
    Stephen Cowan, Hudson Valley Community College
    Ron DiPippo, University of Massachusetts, Dartmouth
    Steve Drouilhet, Sustainable Automation LLC
    Richard Elmer, SUNY Morrisville

Steven Eschbach, FuelCell Energy, Inc.
Leo Evans, Oberlin College
Mark Fillinger, Powerlight Corporation
Erik Foley, Saint Francis University
Jeff Forward, Biomass Energy Resource Center
Randy Gale, California State University
Joel Gilbert, Apogee Interactive
Donna Gold, College of the Atlantic
Bob Grace, Sustainable Energy Advantage, LLC
John Halley, Community Energy, Inc.
Joann Handeland, Cerro Coso Community College
Barry Hilts, University of Pennsylvania
Delaine Hiney, Iowa Lakes Community College
Mike Hupfer, Iowa Lakes Community College
Sam Hummel, Duke University
Ron Jackson, New Jersey Office of Clean Energy
Timothy Jordan, Berea College
Kurt Johnson, U.S. Environmental Protection Agency
Stephen Kalland, North Carolina Solar Center
Ron Kamen, Starfire.net, Inc.
Steven Katona, College of the Atlantic
Patrick Kelly, U.S. Environmental Protection Agency
Julian Keniry, National Wildlife Federation
Michael Koman, University of South Carolina
Jerry Kotas, U.S. Department of Energy
Abigail Krich, Cornell Solar Fund
Joel Levin, California Climate Action Registry
Josh Lynch, Greenpeace
Tim Maker, Biomass Energy Resource Center
Keith Martin, Chadbourne & Park
Michael McCluskey, Austin Energy
Anne Marie McShea, New Jersey Office of Clean Energy
Julie McWilliams, University of Pennsylvania
James Meenagh, John Deere Credit Corp.
Mike Meyers, Aspen Systems
Michael Monroe, South Carolina Department of Heath and Environmental Control
Dr. Cornelius Murphy, Jr., SUNY Syracuse
Mike Nadlock, SUNY Morrisville, NY
Buzz Nelson, University of Nevada, Fairfield
Albert Nunez, Capital Sun Group, Ltd.
Mark Olson, Campus Partners
Ralph Overend, National Renewable Energy Laboratory
Jeffrey Paulson, Jeffrey C. Paulson & Associates
Ron Petty, Concordia University
Bill Prindle, The American Council for an Energy-Efficient Economy
Jane Pulaski, Interstate Renewable Energy Council
Steven Rosenstock, Edison Electric Institute
George Ross, Central Michigan University

Julio Rovi, Cadmus Group
Jennifer Sanguinetti, Keen Engineering
Robert Sauchelli, U.S. Environmental Protection Agency
Walid Shayya, SUNY Morrisville
Georg Shultz, U.S. Department of Agriculture
Marty Silber, energy consultant
Wes Slaymaker, Wes Engineering
Mary Smith, Harvard University
Richard Strong, Carlton College
Randy Swisher, American Wind Energy Association
Ray Tena, Arizona State University
Robert Thornton, International District Energy Association
Timothy Veale, Chevron Energy Solutions Company
Heather Rhodes Weaver, wind power consultant
Ted Williams, New York State Department of Environmental Conservation
Mick Womersley, Unity College
Byron Woodman, Community Energy, Inc.
Diane Zipper, Renewable Northwest Project
Ron Zurek, University of Nevada, Fairfield

And finally, the authors would like to thank the NACUBO staff for their help and dedication to this book, especially Bill Dillon, Donna Klinger, Michele Madia, and David Rupp. We also thank Terry Calhoun, The Society for College and University Planning, and Steve Glazner and Lander Medlin, both of APPA. Thanks also to Ellen Hirzy and Karen Colburn for their editorial and production expertise.

Colleges and universities are saving money and even making money with renewable energy, which includes solar, wind, biomass, geothermal, and hydropower. These higher education institutions have two methods for purchasing renewable energy. They can either build a renewable energy project on or near campus, or they can buy renewable electricity generated by others through a local utility or other supplier. This book examines both approaches and explains how and why growing numbers of higher education institutions are powered by renewable energy. It provides guidance on how to consider the various technologies, ownership options, relationships with utilities, and financing strategies. It includes many examples of institutions that are purchasing or installing renewable energy systems and looks at how they have lowered the cost.

Chapter 1 reviews the motivating factors for higher education institutions to purchase renewable energy. Financial considerations are not always the primary reason, but they increasingly may be. Because most forms of renewable energy do not require fuel deliveries, some green power purchases are not subject to the price increases and price volatility of fossil fuels. In addition, on-site generation during peak electrical demand periods can reduce utility bills. Other motivating factors include the need for a reliable source of electricity during power outages and the desire to reduce carbon emissions and other pollutants. Chapter 1 examines the intangible benefits of renewable energy purchases, including leadership and social responsibility, curriculum and educational value, positive public relations, and improved community relations.

Chapter 2 provides the context for buying renewable energy. Some local state and federal policies and initiatives encourage and sometimes require increased use of renewable energy. This chapter reviews the major policies, such as renewable energy portfolio standards (RPS), which require electric utilities to deliver a minimum percentage of their electricity from renewable energy generators; state executive orders, which require renewable energy use for all state agencies, including state colleges and universities; and net metering laws, which allow renewable energy installations to sell electricity back to utilities at the same price at which users buy it.

Chapter 3 looks at renewable energy sources and their prices. Solar photovoltaic (PV) systems have been the main on-site source for colleges and universities. This chapter reviews PV as well as wind, biomass, and geothermal energy. It does not examine hydropower because few campuses are situated to take advantage of this resource.

Chapter 4 reviews the financial resources that are available to cover or lower the costs of renewable energy. A few options include government grants and rebates, private equity funds, and performance contracts. Costs can be minimized by using energy management savings and other methods. In addition, there are resources from within the campus community, such as student fees and gifts.

Chapter 5 explores the purchase of renewable energy from utilities and renewable energy marketers and brokers. It explains renewable energy certificates (RECs), how they are purchased and certified, and how a fixed price contract can serve as a hedge against volatile or rising conventional electricity prices.

Chapter 6 takes the reader through the process of investing in an on-site renewable energy project, including economic analysis, ownership structures, and utility interconnection issues.

Chapter 7 provides some concluding observations and suggests future directions for purchasing or installing renewable energy that maximize the financial returns to colleges and universities.

WHY BUY RENEWABLE ENERGY?

Colleges and universities throughout the United States are increasingly purchasing clean, renewable energy. As administrators consider these investments, they do so at a time of great economic and policy change. Conventional energy prices—particularly for oil and natural gas—are more and more volatile. The national consensus to reduce U.S. reliance on foreign energy sources is stronger than ever. Energy costs are a significant drain on the nation's economy and balance of payments. Reducing these costs improves economic security by lowering the overall exposure to volatile markets, manufacturing and transportation costs, and capital costs. These concerns, plus the growing concern over global climate change and other environmental impacts of energy production and use, have generated a wave of regional, state, and local policies and initiatives.

Consistent with the heightened national interest in energy issues, colleges and universities are paying more attention to energy. It is being incorporated into environmental studies, engineering, and business curricula. Growing numbers of institutions are pursuing climate initiatives and greenhouse gas inventories based largely on energy use. New buildings are more often green buildings that integrate sophisticated and comprehensive energy-efficiency measures. More institutions are also pursuing the direct use of renewable energy, either by purchasing "green power" at a premium from an electric utility or other supplier or by building renewable energy generation facilities on or near their campuses.

In 2006, more than 200 U.S. colleges and universities are either purchasing renewable energy or producing renewable energy on site for a portion of their energy needs. Many of them are using renewable energy investments to demonstrate leadership and innovation. These institutions are helping move the energy market away from traditional, heavy reliance on coal- and nuclear-based sources of electricity. Renewable energy initiatives have typically engendered strong support from students, faculty, administration, alumni, and the community.

Although there are many reasons to use renewable energy on campuses, this book focuses on the business case. It addresses mainly electricity use but also heating and cooling for campus buildings. It does not address the use of renewable fuels in transportation.

College and university decision makers have several motivations for purchasing green power. The desire to demonstrate leadership is a primary reason, even though renewable energy costs more than traditionally generated electricity at this time. There is a strong national aspiration to become more energy independent in the post September 11 world. Although oil typically is not used to generate electricity in power plants outside of the Northeast, a university can show its commitment to U.S. energy sources by purchasing domestic and renewable power. Renewable energy purchases sometimes result from student-led initiatives with strong support from the faculty. There is anecdotal evidence that environmental stewardship helps with student recruitment and retention. Clean energy purchases improve public relations, and they can help create regional jobs and improve community relations.

In addition to these intangible factors, there are financial considerations that may make spending a little more for green power a beneficial activity. Some aspects of a clean energy purchase can be quantifiable in terms of money saved over time. It can provide a financial hedge against volatile energy prices. An on-site renewable energy system can provide electric backup for critical infrastructure during emergencies and blackouts, when the reliability of systems that transport conventional fuels for backup generation can be severely challenged. In some cases, an on-site generation system can provide income.

▶ TANGIBLE BENEFITS OF PURCHASING RENEWABLE ENERGY

The principal financial benefits of purchasing renewable energy are:

- Hedging against volatile fuel prices
- Reducing peak demand charges
- Increasing reliability
- Qualifying for state implementation plan (SIP) credits
- Producing income from on-site generation
- Earning LEED points

Hedging Against Volatile Fuel Prices

Rising natural gas, oil, and even coal prices are financial drivers to increase the use of renewable energy. The spiking price of heating oil and natural gas has negatively affected the budgets of colleges and universities. Renewable energy purchases and installations can potentially lower the risk by providing stability and long-term pricing. They provide a hedge against the price volatility of traditional oil- and natural gas-generated electricity. Electricity consumption continues to grow worldwide. According to the U.S. Department of Energy, total consumption in the United States is projected to grow at an average annual rate of 1.8 percent through 2025.[1] Colleges and universities have an incentive to diversify their energy sources to lower the financial risk that results from rising prices.

Oil Prices

According to the *Annual Energy Outlook 2005 with Projections to 2025*, issued by the U.S. Department of Energy's Energy Information Administration (EIA), world crude oil prices were at the most recent low of $10.29 per barrel (in 2003 dollars) in December 1998. From that time until December 2001, prices ranged from $20 to $30 per barrel. In January 2004, the price was about $33 per barrel. By October 2004, the price was about $44 per barrel.[2] In early August 2005, oil hit an all time high of $67 per barrel.

The price per barrel of crude oil broke the $70 mark for the first time in late August 2005 in the aftermath of Hurricane Katrina. Many energy experts expect that oil and refined petroleum prices will continue to rise due to severe political instability in oil-producing regions of the world, strong global demand, and constraints on U.S. and world refining capacity.

Oil pressures also increased as a result of Hurricane Katrina, which damaged and idled an estimated 20 percent of the U.S. refining capacity. U.S. refining facilities have no excess capacity, so the oil industry is constrained in the amount of crude oil it can refine, and there have been no new refineries built since 1976. Hurricane Rita's effects on the U.S. energy market began to occur before the storm arrived in September 2005, as oil refineries in the Gulf region were shut down in preparation for the hurricane. These facilities refine almost 25 percent of the nation's gasoline and jet fuel. The shutdowns were temporary, yet they illustrated the precarious nature of the energy market, which was still reeling from Hurricane Katrina.

Energy experts debate when world oil supply will peak. Many experts expect the peak to be within 20 to 25 years. Oil is used much more extensively for transportation than for generation of electricity, except in New England, where it is used for both electricity generation and heating. As oil prices rise, natural gas and coal prices tend to follow.

Natural Gas Prices

The number of power plants that use natural gas for electricity production has increased significantly since 1990, and this trend is expected to continue. The EIA *Annual Energy Outlook 2005* notes that in 2003, 16 percent of electricity was generated through natural gas. It projects this percentage to increase to 24 percent by 2025. While coal use will increase, its share in the mix is projected to decrease slightly, from 51 percent in 2003 to 50 percent in 2025.[3]

Natural gas prices have risen dramatically, from $2 per million Btu in 2000 to $8 per million Btu in August 2005.[4] After Hurricane Katrina, gas prices rose sharply again, to over $14 per million Btu by October 2005.

Natural Gas and Energy Volatility: Report from the American Gas Foundation, prepared for Oak Ridge National Laboratory, found that since 1999 there has been almost no underutilized supply capacity to handle demand resulting from fluctuations in weather. Also, since natural gas plants now generate a higher percentage of electricity, demand has increased and can swing quickly. Since 1997, electricity generation with natural gas increased by more than 62 percent. Natural gas and electricity have had the biggest increase in volatility in the last 10 years, and energy prices overall are more volatile.[5]

Natural Gas Outlook to 2020, another report by the American Gas Foundation, analyzed three possible policy scenarios that would affect natural gas markets through 2020: "expected," "expanded," and "existing." The report says that none of the scenarios will result in the natural gas market that existed through most of the 1980s and 1990s, when surplus supplies and relatively low prices were consistent. Under the "expected" scenario, the report forecasts that natural gas prices will be in the range of $5–$6 per MMBtu for most of the time period and a set nominal price of $8.15 for 2020. It predicts that natural gas supply will be provided by more diverse locations (such as the Alaska pipeline) and sources (such as liquefied natural gas, or LNG). It anticipates that natural gas consumption will continue to increase while demand will rise. Two-thirds of the demand growth is expected to result from electricity generation requirements. According to the report, natural gas demand will be less predictable over the next 15 years.[6]

Coal Prices

The price and price volatility of coal also has increased significantly. From early 2003 to the end of 2004, the price of high-Btu, lower-sulfur eastern coal doubled.[7] Prices have increased from about $25–$28 per ton to $50–$60 per ton in less than two years.[8]

Electricity Prices

According to the Energy Information Administration, the average retail electricity price for commercial customers in June 2005 was 8.30 cents per kWh, an increase from 7.98 per kWh in 2004. Factoring in residential and industrial customers, the 2005 average retail price of electricity from January through June 2005 was 4.2 percent higher than in 2004.[9]

Reducing Peak Demand Charges

Peak load is an important factor is determining an institution's electric bill. Peak load is the maximum load in kilowatts for a facility or group of facilities (if master metered) for a specified time period, often 15–30 minutes. This factor is significant to electric rates because the peak load determines the demand charge. If the demand charge includes a ratchet charge, the ratchet charge can affect the bill for 6–12 months depending on the utility's rate schedules. Therefore, it is important to manage peak load by utilizing load reduction at specific periods of time.

Many utilities offer programs and incentives to encourage peak load reduction. They want to discourage high peak loads, since they need the capacity to serve their customers' maximum electric requirements without building additional capacity for these short periods of time. Utilities with less spare capacity tend to have more incentives. Some utilities provide an incentive if a customer agrees to implement load curtailment when the utility is in danger of brownouts or blackouts. The carrot is the incentives. The stick is high demand charges.

Components of a Utility Bill

By understanding utility rate structures, energy managers can minimize peak demand and have a substantial effect on controlling the overall bill. This information can be obtained from the utility's Web site or account managers. Overall rate class categories include residential, commercial, and industrial. A service class is a subcategory of rate class and determines which tariff applies to the customer. For large campuses, electricity is sometimes delivered to a university substation and then routed to the individual buildings. These buildings may have their own meters. A typical utility bill includes:

- Meter readings: Previous and present periods
- Customer charge: A fixed monthly administrative charge for customers within the rate class
- Demand charge or peak demand: The amount of power in kilowatts used during a specific time. Peak demand charges are common. Because of customer dissatisfaction, some utilities have taken the peak load and ratchet charges and integrated it with other charges.
- Energy charge: Kilowatt-hours of electricity used in a specified time period, based on monthly consumption. Based on the institution's tariff, the energy charge can be billed in various ways:
  - Flat rate: The cost for each kWh is the constant regardless of quantity or time of use.
  - Block rate: The kWh price depends on the quantity consumed. Most often, this is a declining block where an increase in consumption results in lower price per kWh.
  - Time of use or time of day rates: The price varies based on the time of day in which the kWh was consumed. The specific hours of the time periods will be spelled out in the tariff.
  - Time of year rates: Rates are higher in the summer and lower in the winter for most utilities. Typical breakdowns for summer and winter are June to October and November to May.
  - Some utilities break out demand and energy charges based on generation, transmission, and distribution.
- Fuel adjustment charge: Depends on the fuel costs that a utility pays to produce the electricity as compared to the projected fuel costs when the tariffs were developed. The percentage of the bill that the fuel adjustment charge constitutes varies by utility, but it can be substantial. It is based on kWh used by the customer and is either a charge or a credit based on whether fuel costs increased or decreased more than the projections. Customers of utilities that use natural gas and oil for electricity generation are most affected by the fuel adjustment charge.
- System benefit charges (SBC) or public benefit fund: In some states, a small charge applied to energy efficiency or other energy programs. In Oregon, large users can apply the amount of their SBC charges to renewable energy projects.

A few utilities include:

- Ratchet charge or demand ratchet: A percentage of the maximum demand for a specified period of time, for example, a 15-minute block. This charge is applied for an extended period, generally 12 months. If the current month's demand is higher than the ratchet charges, then the charge will be based on the higher demand. If the current month's demand is lower, the customer will still be charged the higher charge. Therefore, if a university has a ratchet charge and reaches a very high demand during a 15-minute period on a hot summer afternoon, it will be charged a percentage (generally higher than 50 percent and up to 100 percent) of that peak demand charge for up to a year. This charge can be very expensive and is the reason why peak load reduction is so important.

Utilities and Demand Response Programs

Edison Electric Institute (EEI), the association of investor-owned electric companies, has a listing of approximately 100 demand response programs provided by utilities throughout the United States.[10] With huge electric demand, several utilities broke their previous records for peak load in the summer of 2005. This unprecedented demand presents vast challenges to the reliability of the nation's transmission and distribution grids. Although there are programs in all regions, there tend to be more incentives to conserve and operate clean on-site projects in states and regions where there are more electricity capacity constraints. Local utilities can provide information on their demand management programs.

How On-Site Generation Helps

The installation of on-site renewable energy systems is a means of reducing peak load. This approach is especially relevant with PV systems with batteries, since peak load is affected most by high air-conditioning loads. There are many other ways to manage loads, including effective energy management and building control systems. Many utilities will contact their largest customers if they are reaching peak to encourage them to decrease their electric consumption on some systems.

For clean distributed generation to reduce peak demand, the renewable energy source needs to match the peak demand. For example, solar would be appropriate for reducing peak during a sunny summer afternoon. In this case, installation of a wind turbine would not be the appropriate choice. The system can be hardwired and sized properly to be able to handle critical functions if desired.

Reliability and Maintaining Critical Functions During Blackout or Emergency

Electric reliability is critical for colleges and universities, as it is for all sectors of the economy. Campuses have a serious need for uninterruptible power because of their range of sophisticated equipment. Ensuring electric reliability for critical campus functions—including medical clinics, computers, security, and sewage treatment—will avoid significant costs that could result from emergencies.

The transmission grids in the United States are aged and under tremendous stress from increased demand. The blackout of August 2003 affected 50 million people from the Midwest to the East and in Canada. According to the Department of Energy, Standard & Poor's estimated that economic losses from the blackout were about $6 billion.[11]

The power outages that resulted from Hurricane Katrina in August 2005 contributed significantly to the devastation in the Gulf region. Universities were not exempt from the crisis. Tulane University, for example, had backup diesel generators for refrigeration of tissue samples from the Bogalusa Heart Study, a long-term epidemiological study on risk factors for children's heart disease that began in 1973. After the power was out for three days and the diesel generators ran out of fuel, the samples were ruined.

Many campuses have on-site generation, usually operated from diesel or gas. Every institution should know how many days of fuel it has stored on site and then consider the consequences of a fuel supply disruption. As recent hurricanes and other emergencies have shown, transportation of these fuels can be delayed.

Diesel backup generators generally operate only in an emergency and thus do not reduce utility bills. The fuel is an extra expense. If a college installs the same capacity in renewable energy (including battery storage), its system will pay for itself over time. Its initial costs are higher than those for a diesel generator, yet unlike the diesel it pays for itself. With no fuel costs, the institution can operate the renewable energy system all the time, not just for backup. Its operation will reduce utility bills. There is some additional operations and maintenance cost (O&M), but it is small.

For backup power, it is necessary to specify which critical functions will need backup power because a renewable energy-distributed generation system will be too expensive to cover the entire load of a given building. The renewable energy system needs to be sized to meet the critical load, which can be put on a separate power circuit. Electricity storage through the use of batteries would meet the anticipated load when the renewable energy system is not generating electricity (for example, solar at night). See chapter 4 for more discussion of battery storage.

SIP and Carbon Credits

States that are not in compliance with the Clean Air Act must submit a state implementation plan (SIP) to the Environmental Protection Agency outlining plans for coming into compliance. Public colleges and universities may be able to help states meet this requirement through renewable purchases or energy management projects. In return, a state may be willing to help fund on-site generation, such as a wind turbine or solar installation. Montgomery County, Maryland, purchased wind power, which is being credited to the SIP. To the authors' knowledge, state and local SIP funds have not been used to finance renewable energy or energy-efficiency measures at colleges or universities, but it is possible.

Income from On-Site Generation

Carleton College provides an example of how the installation of a wind turbine can yield income. Carleton borrowed from its endowment for construction of its turbine. It is selling the output to its utility and receiving a Minnesota Tax Production Credit. This income more than covers the cost of repayment to the endowment (see the case study in chapter 3).

Renewable Energy for LEED Points

Both grid purchases and on-site renewable energy are points in the U.S. Green Building Council's Leadership in Energy and Environmental Design (LEED) Green Building Rating System. Many colleges and universities are building LEED facilities. Some states are requiring LEED for new state buildings, including public colleges and universities. It is worth examining the cost-effectiveness of on-site generation or a grid purchase for this purpose.

For example, the Los Angeles Community College District (LACCD), the University of California system, and the California State University system require LEED or LEED-equivalent for all new buildings. Under Proposition A/AA, LACCD requires that all new buildings get LEED silver certified. LACCD has nine colleges and more than 120,000 students.[12]

"Energy and Atmosphere" count for 17 possible LEED points. For on-site generation, one to three points are possible, depending on the percentage of a building's electric load that is met by the system. For purchasing green power, a facility can receive one point if at least 35 percent of the building's electricity is from renewable energy for at least two years. For certification, a building must achieve the follow number of points:

- Certified: 26–32
- Silver: 33–38
- Gold: 39–51
- Platinum: 52–69[13]

▶ INTANGIBLE BENEFITS OF PURCHASING RENEWABLE ENERGY

Baruch Lev defines intangible assets as "sources of future benefits that lack a physical embodiment." He cites research indicating that the U.S. corporate investment in intangibles in 2000 was $1 trillion. The U.S. manufacturing sector invested $1.1 trillion in physical assets the same year. Investments in intangibles will soon exceed investments in tangible assets.[14] Intangible or "soft assets," Lev says, are responsible for more than one-half of the market capitalization of public companies. These intangibles account for most corporate growth and shareholder value.[15]

As physical assets have become commodities, intangible assets increasingly determine the performance of U.S. and world corporations. Some financial analysts contend that a corporation's superior environmental performance is an indicator of overall excellent management and, in turn, improves shareholder value.

Jonathan Low and Pam Cohen Kalafut report that one-third of an organization's value results from intangible assets, and 35 percent of portfolio managers' decisions on where to allocate dollars for investment are based on intangibles and nonfinancial information.[16] They identify reputation, leadership, brand equity, adaptability, and innovation as among the 12 most important value indicators.[17]

A report by Global Environmental Management Initiative (GEMI), a nonprofit organization whose members are major U.S. corporations, analyzes the relationship between shareholder value and environment, health, and safety (EHS) performance. The report states that "50 to 90 percent of a firm's market value can be attributed to intangibles like EHS and that 81 percent of Global 500 executives rate EHS issues among the top ten driving value in their businesses."[18]

The most relevant indices are dependent on the specific industry. According to GEMI, some of the more important measures include leadership and strategy, brand equity, and environmental and social reporting. The report discusses sample, leading indicators for each key intangible. For example, sample performance indicators for leadership and strategy include "commitment to EHS/sustainability principles and goals" and "articulation and execution of EHS strategy." A sample performance indicator for brand equity is "perception of brand as environmentally and socially responsible."[19]

The subject of intangible value has been studied in regard to participation in Energy Star, which focuses on improving the energy performance of institutions, corporations, and individuals. Lou Nadeau of the Eastern Research Group has studied the intangible value of participation as it relates to corporate reputation. According to Nadeau's research, real estate investment trusts (REITs) that were involved in Energy Star partnerships received a value that represented 1.6 percent of assets and was worth 3.7 percent of these REITs' total market value.[20]

The intangible value of a college or university's commitment to renewable energy is hard to measure, yet it is significant. It is difficult to perform a return-on-investment analysis for these nonfinancial performance indicators, which include:

- Leadership and social responsibility
- Curriculum and educational value
- Student, faculty, and alumni support
- Student recruitment and attitudes
- Public relations
- Collaborations and partnerships
- Positive community relationships

Leadership and Social Responsibility

Many college and university administrators consider environmental stewardship as integral to their mission and the "right thing to do." In the view of various constituents, clean energy purchases and installations support the mission of universities as leaders.

Many people view global climate change as one of the most pressing problems of the coming decades. Most scientists conclude that global warming due to human activity is already affecting climate patterns throughout the world. These issues are influencing international relations, and they are certain to gain importance as the effects of global warming become even more pronounced.

Energy usage is a major contributor to the changes in concentrations of greenhouse gases (GHG) that raise the risk of global climate change. Many colleges and universities study the link between energy and global warming with the goal of finding methods to mitigate its effects. The American public is increasingly aware of global warming and the greenhouse effect as a result of Hurricane Katrina, the passage of the Kyoto Protocol, and extensive media coverage. As colleges and universities educate and train tomorrow's leaders, the issue of "walking the talk" in regard to energy use is of paramount concern.

Administrators from a growing number of colleges have demonstrated their strong leadership by committing to 100 percent clean electricity for their campuses. The first

CASE STUDY

New Jersey Higher Education Partnership for Sustainability Providing Leadership

In a unique arrangement, the colleges and universities of New Jersey have undertaken a collaborative effort to address climate change, including the use of renewable energy sources. In 2001, the presidents of all 56 colleges and universities in the state signed on to the New Jersey Sustainability Greenhouse Gas Action Plan. The implementation of the plan, coordinated by the New Jersey Higher Education Partnership for Sustainability (NJHEPS), has involved the following steps:

- Energy reports and action plans for reducing emissions and costs, created in PowerPoint format for presentation to college decision makers
- Appointment of sustainability coordinators from facilities and faculty to implement the plan
- *High Performance Campus Design Handbook*, guidelines for green design
- The Building Energy Audit and Technical Assistance Initiative, a database providing baseline building energy information to use for benchmarking and goal setting in the entire sector
- Establishment of emissions and energy reduction targets at each campus
- *The Renewable Energy and Global Warming Curriculum Module*, a Web-based problem-solving product
- Implementation of emissions and energy reductions at multiple campuses through energy-efficiency improvements and renewable energy investments

Renewable energy has been an important component of the action plans, and there has been a surge of solar energy investment led by Kean University, Ramapo College, New Jersey Institute of Technology, and Monmouth University.

NJHEPS is the coordinating mechanism for the colleges and universities and fosters sharing of information, presentation of educational materials, and publication of technical information. The organization produced the *Campus Energy Toolkit: Tips, Strategies and Case Studies to Reduce Energy Costs and Emissions at New Jersey Colleges and Universities*, which will be useful to colleges and universities in other states as well.[21]

Key to the implementation of the solar projects and the broader action plans has been the development of a sustainable funding base. NJHEPS works with individual campuses to develop sustainable business plans with increased support from the higher education institutions they serve, corporations, businesses, government, and foundations, as well as from fees for services. For more information on NJHEPS and its activities, see www.njheps.org.

higher education institution in the country to contract for 100 percent wind energy was Concordia University in Austin, Texas. College of the Atlantic in Bar Harbor, Maine, was the first to contract for a 20-year purchase for 100 percent wind (although a change of supplier required that the contract be shortened to 10 years). Other colleges and universities that are purchasing 100 percent clean energy include Unity College and Colby College, both in Maine, and Western Washington University.

For colleges and universities that have had active recycling and energy conservation programs throughout the years, a renewable energy purchase is often viewed as the natural next important step that augments other environmental initiatives. A green power purchase is one way to raise awareness of an institution's commitment to action on environmental issues.

FIGURE 1-1: SOLAR INSTALLATIONS IN NEW JERSEY HIGHER EDUCATION

Kean University	28.0 kW
Monmouth University	454.0 kW
Montclair State University (demonstration)	< 5.0 kW
New Jersey Institute of Technology	50.0 kW
Ramapo College	100.0 kW
Richard Stockton State College	18.0 kW
Stevens Institute of Technology	125 kW
University of Medicine and Dentistry	2.5 kW

Source: New Jersey Higher Education Partnership for Sustainability

Curriculum and Educational Value

A clean energy purchase provides opportunities for curriculum and educational value. Colleges and universities tie their renewable energy initiatives to environmental and energy-related curriculum for engineering, public policy, architecture, economics, business, and other disciplines.

Iowa Lakes Community College in rural northwest Iowa has a new two-year degree in wind energy and turbine technology. This program prepares students for a career in the fast-growing wind industry, which has an employment demand. Using the college's wind turbine as a training lab, the program focuses on construction, maintenance, repair, and operations. Courses include field training, construction and site location, wind turbine operations, mechanical systems, and power generation and distribution.[22]

Student, Faculty, and Alumni Support

Clean energy purchases are popular within the college community. Students, faculty, administrators, and alumni at many institutions are very supportive of spending a little more money to procure green power. Often, motivated students initiate the idea, perform the initial research, and then gain the support of the faculty and administration. In the end, it is a collaborative effort that can strengthen the campus community.

A 2001 report by the National Wildlife Federation—*State of the Campus Environment: A National Report Card on Environmental Performance and Sustainability in Higher Education*—is based on responses from more than 1,000 presidents, provosts, and chiefs of administration and operations at 891 (almost 22 percent) of the colleges and universities in the United States. The survey analyzed why higher education institutions develop and implement environmental programs and reported the following reasons:

- Feel environmental programs fit in with the culture and values of the campus (64 percent)

- Have found environmental programs are good public relations (47 percent)
- Have found them to be cost-effective (40 percent)
- Have found environmental programs help recruit students (17 percent)[23]

Student Recruitment and Attitudes

Many factors contribute to prospective students' choice of a college. An institution's sustainable activities and reputation may be one factor. In a *Wall Street Journal*/Harris Interactive study that ranked the nation's top business schools, recruiters listed the three "most impressive features." For number one-rated Tuck School of Business at Dartmouth College and number five-rated Yale University School of Management, recruiters cited "social responsibility" as a most impressive feature.[24] World Resources Institute ranks business schools on their sustainability curriculum in preparing future business leaders as part of its Beyond Grey Pinstripes program.

Student attitudes toward renewable energy can be gleaned from a review of student votes on green power initiatives. On many campuses, students overwhelmingly endorsed clean energy purchases even if they had to pay an additional fee. If students are willing to pay, the support is strong. For a few dollars per student, many administrators feel that this is an inexpensive way to make students know that the administration takes its environmental commitment seriously.

An opinion poll sponsored in 2003 by the Minnesota Public Interest Research Group (MPIRG) surveyed students' attitudes toward energy:

> "There was strong support for increasing renewable energy efforts in Minnesota, even if it means paying more each month. Over 78 percent of students expressed support for making the state's current "Renewable Energy Objective" into a mandatory standard that would require power companies to produce 10 percent of their electricity from renewable sources. Two-thirds want renewable energy to be the highest priority for meeting future energy demand. Almost 37 percent of those surveyed were willing to pay $6 to $15 per month more on their electricity bills to increase renewable energy and phase out nuclear power. An additional 20 percent were willing to pay $1 to $5 more per month."[25]

Just as recycling gained support 20 years ago and is now commonplace, renewable energy is quickly gaining popularity. A subset of students expects that their institution buys or produces clean energy.

Public Relations

Colleges and universities have received extensive, positive coverage of their clean energy initiatives in print, broadcast, cable, and online media. Institutions have received national awards for their commitments and leadership in stimulating the green power market.

News articles, TV news reports, and radio stories generally have more credibility with the public than advertisements. People also tend to read articles more than advertisements. Green power purchases offer a substantial opportunity for a positive media presence, with significant benefits in terms of improved reputation and exposure.

For example, the story of the University of Pennsylvania's wind purchases (see the case study in chapter 4) was picked up by the Associated Press, several major newspapers, and radio and TV stations. For a story that aired on a local television news show, a firm hired by the university determined that the 30-second advertising equivalency was $3,200.

To maximize the impact of a clean energy purchase or installation, it is important to publicize it so that students, faculty, community members, and others are aware of the commitment. A few years ago, a wind purchase could more easily garner significant news coverage on TV, radio, and in the press. As clean energy purchases become more common, the bar has been raised and it is increasingly difficult to get the same level of exposure. EPA's Green Power Partnership has a host of tools to help. Coordination among facilities, administration, business officers, students, and the public affairs department can also optimize the exposure. It is also important to promote the college or university's commitment to renewable energy in its admissions materials—especially if the institution has related curriculum. On-site generation facilities can be featured in tours for prospective students and their parents.

Collaborations and Partnerships

Progressively more colleges and universities, corporations, and utilities are forming collaborations and partnerships to advance renewable energy and mitigate global climate change. In September 2005, Duke Energy of Durham, North Carolina, entered into collaboration with Duke University for a Climate Change Policy Partnership to develop policy options that address global climate change. Duke Energy pledged $2.5 million to the university in support of this new partnership. The partnership will also include other corporate and academic partners.[26] Southern Company of Atlanta, one of the nation's largest utilities, has partnered with Georgia Institute of Technology to determine the feasibilty of an offshore wind power project that would be the first project of this type in the Southeast.[27]

Community Relationships

Clean energy purchases stimulate the green power market and encourage green power developers to build capacity and increase supply. The higher education market has had a direct, positive impact on green power development. An important example results from the commitment of Carnegie Mellon University, Penn State University, and the University of Pennsylvania. After these institutions purchased two-thirds of the output of a wind farm in Somerset, Pennsylvania, demand for wind increased and additional wind farms were built in the state. In 2001, Carnegie Mellon bought 5 percent wind, the equivalent of 1 turbine; Penn State purchased 5 percent wind, the equivalent of 3.5

turbines; and the University of Pennsylvania purchased 5 percent wind, the equivalent of 5 turbines. In 2003, Penn increased its commitment to 10 percent of its load, or 10 turbines. More than 40 colleges and universities in Pennsylvania are now purchasing wind power generated by the six utility-scale wind farms in the state.

According to American Wind Energy Association, Pennsylvania Governor Edward G. Rendell announced in early 2005 that an international wind turbine maker plans to build a manufacturing plant and locate its U.S. headquarters in Pennsylvania. The new plant is expected to result in 1,000 new jobs in the state over the next five years.[28]

ENDNOTES

1. Energy Information Administration, U.S. Department of Energy, *Annual Energy Outlook 2005* (Washington, DC: Energy Information Administration, 2005), p. 4.

2. Ibid., p. 2.

3. Ibid., p. 6.

4. www.nytimes.com/2005/09/28/business/28chemical.html (accessed October 4, 2005).

5. Bruce Henning, Michael Sloan, and Maria de Leon, *Natural Gas and Energy Price Volatility* (Washington, DC: American Gas Foundation, 2003), pp. 2-3; available at: www.gasfoundation.org/ResearchStudies/volatility.htm.

6. Paul Wilkinson, Chris McGill, Kevin Petak, and Bruce Henning, *Natural Gas Outlook to 2020: The U.S. Natural Gas Market—Outlook and Options for the Future* (Washington, DC: American Gas Foundation, 2005).

7. www.platts.com/Coal/Resources/News percent20Features/coal_prices_2005 (accessed October 3, 2005).

8. www.washingtonpost.com/ac2/wp-dyn/A42157-2005Apr10?language=printer (accessed May 15, 2005).

9. www.eia.doe.gov/cneaf/electricity/epm/epm_sum.html (accessed October 17, 2005).

10. Steven Rosenstock, *EEI Member and Non-Member Residential/Commercial/Industrial Efficiency and Demand Response Program for 2005,* available at: www.eei.org/industry_issues/retail_services_and_delivery/wise_energy_use/programs_and_incentives/progs.pdf Washington, DC: Edison Electric Institute (accessed July 25, 2005).

11. www.electricity.doe.gov/about/boxstory1.cfm?section=about&level2=box1 (accessed August 11, 2005).

12. www.laccd.edu/about_us/fast_facts.htm (accessed November 10, 2005).

13. Green Building Rating System for New Construction & Major Renovations, Version 2.2, 2005, available at: https://www.usgbc.org/FileHandling/show_general_file.asp?DocumentID=1095 (accessed November 7, 2005).

14. Baruch Lev, "Intangible Assets: Concepts and Measurements," in *Encyclopedia of Social Measurement,* vol. 2 (2005), p. 299.

15. Baruch Lev, "Sharpening the Intangibles Edge," *Harvard Business Review*, June 2004.

16. Jonathan Low and Pamela Cohen Kalafut, *Invisible Advantage, How Intangibles are Driving Business Performance* (Cambridge, MA: Perseus Press, 2002), p. 7.

17. Ibid., p. 14.

18. *Clear Advantage: Building Shareholder Value* (Washington, DC: Global Environmental Management Initiative, 2004), p. 1; available at: www.gemi.org/GEMI percent20Clear percent20Advantage.pdf. (accessed October 19, 2005).

19. Ibid., p. 10.

20. Lou Nadeau, *Participation in Voluntary Programs: Corporate Reputation and Intangible Value*, available at: www.nrel.gov/analysis/seminar/docs/2005/ea_seminar_aug_11.ppt (accessed August 31, 2005).

21. See www.njheps.org/toolkit.pdf.

22. www.iowalakes.edu/registrar/catalog/WT_wind_energy_turbine.htm (accessed September 22, 2005).

23. Mary McIntosh, *State of the Campus Environment: A National Report Card on Environmental Performance and Sustainability in Higher Education* (Washington, DC: National Wildlife Federation, 2001), p. 23.

24. *Wall Street Journal*, September 21, 2005, p. R6

25. www.mpirg.org/reports/EnergySurveyReport.pdf (accessed June 20, 2005).

26. www.dukenews.duke.edu/2005/09/dukeenergy.html (accessed September 27, 2005).

27. www.gatech.edu/news-room/release.php?id=569 (accessed November 1, 2005).

28. www.awea.org/news/news050426qmk.html (accessed October 4, 2005).

ENERGY PROCUREMENT IN A CHANGING WORLD

There are a number of policy drivers for renewable energy investments and purchases. The economics of renewable energy are affected by local, state, and federal policies. State and local governments take several approaches to renewable energy policy:

- *Renewable portfolio standards* (RPS) require electric utilities to purchase and supply renewable energy.

- *Executive orders* require states' public institutions to use renewable energy.

- *Net metering laws* require utilities to purchase renewable energy generated by their customers.

- *Local initiatives* require local governments to procure renewable energy.

- *Financial incentives* come in the form of grants, rebates, and tax incentives.

In addition to these financial drivers and investment requirements, a number of voluntary initiatives encourage or commit institutions to renewable energy investment at the regional and national levels. These initiatives include:

- U.S. Mayors Climate Protection Agreement

- Northeast Governors and Eastern Canadian Premiers Agreement on Climate Change

- West Coast Governors' Global Warming Initiative

- Western Governors' Association Clean and Diversified Energy Initiative

- Regional Greenhouse Gas Initiative

- Chicago Climate Exchange

Two federal laws have implications for renewable energy investment:

- the Energy Policy Act of 2005, which includes financial incentives and addresses net metering
- the Clean Air Act, which allows renewable energy investments to qualify for clean air credits.

▶ RENEWABLE PORTFOLIO STANDARDS

Renewable portfolio standard (RPS) legislation requires electric utilities to include a certain portion of renewable energy in their electricity mix. Twenty-two states and the District of Columbia currently have RPS. As of this writing, there is no federal RPS, but many experts expect that this legislation will be reintroduced at the federal level.

Some states use a two-tiered system to define sources of renewable energy and set percentage requirements. Tier one typically includes noncombustion sources such as solar and wind but may also include some forms of biomass. Tier two typically includes combustion sources such as landfill methane, municipal solid waste, poultry-litter incineration, and even coal mine methane.

RPS legislation exists in Arizona, California, Colorado, Connecticut, Delaware, District of Columbia, Hawaii, Illinois, Iowa, Maine, Maryland, Massachusetts, Minnesota, Montana, Nevada, New Jersey, New Mexico, New York, Pennsylvania, Rhode Island, Texas, Vermont, and Wisconsin. A number of states, including New York and California, have accelerated their timetables for implementation.

▶ EXECUTIVE ORDERS AND OTHER STATE-LEVEL AGREEMENTS

Some governors have issued executive orders and renewable energy purchasing commitments for state buildings as well as state colleges and universities. These directives deal with increasing energy efficiency and using renewable energy in state-owned and -operated facilities.

Arizona

Executive Order 2005–05, February 2005. Requires all new state buildings to have 10 percent renewables from RECs or on-site generation.[1]

Connecticut

Executive Order 32, April 2004. Directs state government and universities to increase the percentage of renewable energy to 10 percent by 2010, 50 percent by 2020, and 100 percent by 2050.[2]

Illinois

Executive Order 6, April 2002, Renewable Energy Executive Order for State Government. Directs affected state agencies to purchase at least 5 percent renewable energy for electricity requirements for state-owned or -operated buildings by 2010 and at least 15 percent by 2020.[3]

Iowa

Executive Order 41, April 2005. Requires state agencies to increase renewable energy use and purchase at least 10 percent renewable energy by 2010. Agencies can generate their own renewable energy or purchase it through their utility.[4]

Maine

Executive Order ME PUC 65.407. Ch. 311. 2003. Sets goal for state government to buy at least 50 percent of its electricity from "reasonably priced renewable power sources." These purchases are to be paid for by energy management savings in state buildings, including colleges and universities.[5]

Maryland

Executive Order 01.01.2001.02, January 2001, Sustaining Maryland's Future with Clean Power, Green Buildings and Energy Efficiency. Sets a clean energy procurement goal of 6 percent to be generated from green energy for state-owned facilities.[6]

New Jersey

A group of New Jersey state agencies is meeting about 15 percent of electric needs with renewable energy. The renewable energy contracts include 11 state universities.[7] In addition, New Jersey has a Sustainability Greenhouse Gas Action Plan for reducing emissions by 7.5 percent below 1990 baseline levels by the end of 2005. Results are expected in Spring 2006. Presidents of all of New Jersey's colleges and universities have signed the covenant.

New York

Executive Order 111, 2001, Directing State Agencies to Be More Energy Efficient and Environmentally Aware. Requires agencies to purchase 10 percent renewable energy by 2005 and 20 percent by 2010. They must also implement energy-efficient practices and follow green building standards during new construction or substantial renovation. In his 2003 State of the State Address, Governor George Pataki directed the Public Service Commission to implement an RPS to generate at least 25 percent of the electricity purchased in the state from renewable resources by 2013. This order applies to state universities and colleges, state buildings, and quasi-independent organizations.[8]

Pennsylvania

In October 2004 Pennsylvania doubled its green electricity purchase for state government facilities from 5 percent to 10 percent.[9]

Rhode Island

The state will purchase RECs for 100 percent of the annual energy consumption of the Rhode Island State House.[10]

Texas

In August 2005, Texas more than doubled its RPS to require that approximately 5 percent of the state's energy needs come from renewable sources by 2015 and 10 percent by 2025. The original RPS, established in 1999, mandated that 2,000 MW of new renewables were to be installed in the state by 2009.

▶ NET METERING

Net metering laws enacted at the state level have been a powerful policy driver for renewable energy development. In net metering, electric meters run in both directions, and customers are credited for electricity that they generate in excess of their consumption.

Net metering laws, enacted in 40 states plus Washington, D.C., favor small generators like college and university renewable energy projects because they require the utility to buy the customer's power *at the same price* as the customer buys it from them. Without net metering laws, the college or university can still sell electricity to the utilities, but at a lower price. Utilities justify this practice because they usually buy power at a wholesale price and sell it at a retail price. Furthermore, the wholesale price they offer will usually be lower than the price they offer conventional power plants because the college or university's volume of power output is lower; because it may be generating power with an intermittent, unreliable source like solar or wind; or because the utility is not confident that the college or university will operate and maintain its generator properly so that it will provide power when the utility needs it.

States that allow net metering also place two basic limitations on it: the kilowatt capacity (or kilowatt-hours output) of the generation technology and the generation technology being used. Basically, net metering is restricted to small solar and wind projects. That constraint might be acceptable if a college or university is thinking of installing a small system, but many want to consider substantial systems that supply all or a significant portion of their electricity. If the system will exceed the limitations in the state law, then the college or university must negotiate a price with the utility if it is interested in selling any of the electricity.

State net metering laws vary widely (fig. 2-1). In California, a project sponsor can have up to a 1 MW system, while next door in Arizona the sponsor is limited to 10 kW (one one-hundredth of the California limit). Some states have one limit for solar and another limit for wind. Some have limits on the size of the renewable energy facility (in kW) and others have limits on the amount of electricity that can be net metered from such facilities (in kWh).

If a college or university is located in a state with a net metering law, the local utility will be more interested in purchasing its renewable energy output than a utility in a non-RPS state. While having an RPS increases utility purchases of renewable energy, it does not mean utilities will buy electricity from every renewable energy generator. Generally, they will be less interested in buying from small generators and prefer agreements with a few large generators, such as privately owned wind farms. Even if the utility is interested in the renewable energy from a college or university, it will seek to buy it at the lowest possible cost, and the institution could find itself competing with other generators. Current RPS allocations are relatively small, often requiring 5–7 percent of the utility's electricity to come from renewable energy. The utilities may be able to meet the requirement with just a few large purchases.

FIGURE 2-1: NET METERING RULES

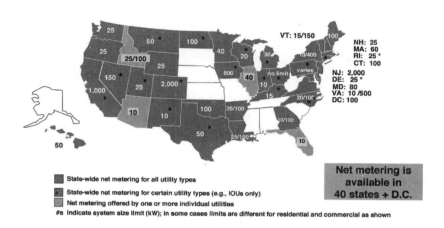

Source: DSIRE Database and the Green Power Network, U.S. Department of Energy

STATE INCENTIVES

State governments have taken a leadership role in providing direct financial support for renewable energy projects. Several states offer some form of assistance, usually through grants or rebates. Many have tax incentives, and some give property tax breaks, which are not directly beneficial to nontaxpaying institutions. (See chapter 4 for more details on the state incentives most applicable to colleges and universities.)

LOCAL INITIATIVES

There has been significant activity on renewable energy initiatives at the city and county levels. Several towns, cities, and counties are purchasing substantial percentages of clean energy through RECs and/or installing clean energy generation. Many have more ambitious goals for the next decade and beyond. A few municipal governments—including Columbia, Missouri, and Austin, Texas—have their own RPS legislation. The Database of State Incentives for Renewable Energy (DSIRE) provides these examples:

- In Madison, Wisconsin, the goal is 10 percent renewables by 2006 and 20 percent by 2010.

- The Los Angeles city council voted in 2001 to procure 14 percent new renewables for city facilities. In 2004, the council voted to increase renewables purchase to 20 percent.

- Boulder, Colorado, purchases wind for municipal buildings.

- Portland, Oregon, purchases 12 percent clean energy with a goal of 100 percent.

- Chicago, Illinois signed an agreement in 2001 to purchase 20 percent by 2005. The city purchased 10 percent renewables in 2003, including landfill gas, wind, and solar power.

- Aspen, Colorado, has a goal of 75 percent green power by 2010.

- San Diego, California passed a green power purchase resolution in 2003 to install 50 MW of renewable energy projects in the next decade.

- Santa Monica, California began purchasing 100 percent clean energy in 1999. It was the first city in the nation to commit to 100 percent green power.

- San Francisco, California voters passed a $100 million bond initiative in 2001 to enable the city to pay for clean energy and energy-efficiency projects, including solar panels and wind turbines for public buildings. The bond is paid down from energy savings, so there is no additional cost to the taxpayers.[11]

Other purchasers include Scottsdale, Arizona; Seattle; Davis, California; Conway, South Carolina; and Myrtle Beach, South Carolina.[12]

Local governments have adopted aggregate purchasing to lower the cost of green power. Montgomery County, Maryland, and surrounding counties aggregate wind purchase. Six county agencies and 11 municipalities in Prince Georges County, Maryland, participate in a bulk purchase. Neighboring Fairfax and Arlington Counties in Virginia aggregate through the Virginia Energy Purchasing Governmental Association for their wind purchase. The Northeast Ohio Public Energy Council is an aggregation of 350,000 residential customers from more than 100 communities. Unless customers opt out, they receive electricity that includes 2 percent renewables.

▶ REGIONAL AND NATIONAL INITIATIVES

U.S. Mayors Climate Protection Agreement

In March 2005, 10 mayors wrote to more than 400 of their colleagues urging joint local action to mitigate global warming. One month earlier, the Kyoto Protocol had taken effect in the 141 countries that had ratified it. The Kyoto Protocol is an amendment to the United States Framework Convention on Climate Control. Countries that ratify this protocol commit to reduce their emissions of carbon dioxide and five other greenhouse gases, or engage in emissions trading if they maintain or increase emissions of these gases. The U.S. Conference of Mayors unanimously approved the Mayors Climate Protection Agreement in June 2005. By April 2006, more than 220 mayors had endorsed the agreement.

Under the agreement, which is sponsored by Mayor Greg Nickels of Seattle, participating cities commit to take three actions:

- Strive to meet or beat the Kyoto Protocol targets through actions ranging from anti-sprawl land-use policies to urban forest restoration projects to public information campaigns.

- Urge the state and federal governments to enact policies and programs to meet or beat the greenhouse gas emission reduction target suggested for the United States in the Kyoto Protocol: 7 percent reduction from 1990 levels by 2012.

- Urge the U.S. Congress to pass the bipartisan Climate Stewardship Act, which would establish a national emission trading system to "increase the use of clean, alternative energy by, for example, investing in "green tags," advocating for the development of renewable energy resources, and recovering landfill methane for energy production."[13]

Northeast Governors and Eastern Canadian Premiers Agreement on Climate Change

This climate change action plan began in 2001 and includes six states and six provinces. The goal of the agreement is to reduce greenhouse gas emissions to 10 percent below 1990 levels by 2020. Colleges and universities in these areas are working in conjunction with states and nonprofits to make progress toward achieving this goal.[14]

West Coast Governors' Global Warming Initiative

Governors of California, Oregon, and Washington signed an agreement in November 2004 to work together to mitigate global warming and reduce greenhouse gas emissions. The agreement focuses on goals and strategies for increasing retail energy sales from renewable resources by 1 percent or more each year through 2015. The governors are also pushing for regional progress to increase the markets for energy efficiency and renewable resources.[15]

Western Governors' Association

In 2004, governors from 18 states agreed to develop 30,000 MW of clean energy by 2015. In addition, the policy resolution urged an increase in energy efficiency of 20 percent by 2020.[16]

Regional Greenhouse Gas Initiative (RGGI)

RGGI is a seven-state initiative, announced in December 2005, to create a regional cap-and-trade program to control carbon dioxide emissions from power plants in the Northeast and Mid-Atlantic. The participating states are Connecticut, Delaware, Maine, New Hampshire, New Jersey, New York, and Vermont. Maryland, Massachusetts, Pennsylvania, Rhode Island, the District of Columbia, the Eastern Canadian Provinces, and New Brunswick are observing the process.[17]

Chicago Climate Exchange

Chicago Climate Exchange (CCX) is a voluntary trading program for carbon dioxide and other greenhouse gases. Partners make commitments to emission reduction targets and timetables. An institution or company that achieves further reductions can trade with a member that does not reach its emission reduction goals. An institution can use offset projects such as a renewable energy installation.[18] When the market further develops, this program will be a potential source of income to the institutions. Tufts University, University of Iowa, University of Oklahoma, University of Minnesota, and Presidio World College are CCX members.

▶ FEDERAL INITIATIVES

Energy Policy Act of 2005

The Energy Policy Act of 2005, which was developed and debated over four years, is the first major federal energy legislation since the Energy Policy Act of 1992. The law includes significant federal tax incentives to increase energy efficiency and the use of renewable energy. The tax incentives went into effect on January 1, 2006.

The act does not include a federal renewable portfolio standard. However, it requires the federal government to purchase 3 percent of its electricity from renewable sources for its facilities by fiscal year 2007, 5 percent by 2010, and 7.5 percent by 2013 and thereafter.

Section 1301 of the act provides a renewable energy production incentive for developers. Qualifying facilities include wind, closed loop biomass, geothermal, landfill gas, and a few other types that are placed in service by December 31, 2007. The credit is based on an amount per kilowatt-hour produced over a 10-year period. For example, the production tax credit for wind facilities is 1.9 cents per kWh. The tax credit is expected to spur the development and increase the supply of renewable energy facilities. A further tax incentive is provided by section 1337. For the commercial installation of solar systems, the business solar investment tax credit has increased to 30 percent from 10 percent for period before January 1, 2008.

The act requires reductions in annual energy consumption in federal buildings and increases requirements for the procurement of energy-efficiency products. It authorizes the expenditure of up to $250 million over five years for rebates on energy-efficient appliances.[19] For new commercial buildings that reduce annual energy and power consumption by 50 percent compared to the American Society of Heating, Refrigerating, and Air Conditioning Engineers (ASHRAE) standards, there is a tax deduction with the maximum of $1.80 per square foot of the building.[20]

Clean Air Act

Renewable energy investments can have a positive impact on local air quality because they reduce reliance on polluting fossil fuel power plants. The federal Clean Air Act requires state and county governments to comply with federal emissions standards or risk losing federal transportation dollars. States have specialized funds to invest in pollution reduction measures. EPA estimates that state and local governments spend $19 billion each year on Clean Air Act compliance measures and will be spending $27 billion by 2010. There may be opportunities for colleges and universities to work with their states to receive state funding for renewable energy investments in order to help states comply with the Clean Air Act (see chapter 4).

ENERGY EFFICIENCY: THE FIRST STEP

Consumers who wish to buy a portion of their electricity from renewable energy typically pay a premium, while on-site renewable energy systems frequently have high first costs and long paybacks. Thus, colleges and universities that have invested in renewable energy usually have invested in energy efficiency first or have invested in both at the same time. One of the reasons for making a sequential investment is that the cost savings from the efficiency measures can provide funds for the renewable energy purchase. This technique is a common one for financing renewable energy purchases and has been used by the University of Pennsylvania, University at Buffalo, SUNY, Catholic University, and Concordia University.

In most cases, energy efficiency is a more cost-effective investment than renewable energy. Before purchasing renewable energy or installing on-site generation, it is prudent for an institution to first invest in energy-efficiency upgrade projects throughout campus facilities. A range of measures—such as thermal insulation, high efficiency lighting systems, and appliances—have some of the highest financial returns of any investment a higher education institution can make in its physical plant. It makes little economic sense to build a solar power system to provide electricity to a building that is going to waste the electricity with inefficient equipment. Also, by investing in energy efficiency first, the college or university lowers its electricity bill. When it then purchases green power, the percentage will be lower, and the green power bill will be less.

An on-site renewable energy project can be a prudent investment, depending on how it is structured, what state financial incentives are available, and what legal agreements the college or university can reach with its local utility. When major energy-efficiency capital investments are being pursued, it may make sense for financing purposes to include a contemplated on-site renewable energy project with them as part of a total package. Under this scenario, the long payback and relatively low internal rate of return (IRR) of the renewable energy project, combined with the short paybacks and high IRRs of the energy-efficiency investments, would yield an overall payback and IRR that is more acceptable than if the renewable energy project were appraised on its own. A renewable purchase can be paid without up-front costs out of the savings if performance contracts are used to finance the efficiency projects and the term of the contract is extended slightly (see chapter 4.)

ENDNOTES

1.   www.governor.state.az.us/eo/2005_05.pdf (accessed July 5, 2005).

2.   www.smartpower.org/executiveorder.html (accessed July 5, 2005).

3.   www.epa.state.il.us/green-illinois/executive-orders/number-6-2002.html (accessed July 11, 2005).

4.   www.governor.state.ia.us/news/2005/april/april2205_1.html (accessed July 11, 2005).

5.   www.state.me.us/governor/baldacci/vision/environment.html (accessed July 14, 2005).

6.   www.dsireusa.org/documents/Incentives/MD07R.htm (accessed July 11, 2005).

7.   www.state.nj.us/dep/dsr/bscit/CleanEnergyMain.htm (accessed July 13, 2005).

8.   Governor George E. Pataki, State of the State Address, January 8, 2003, available at: www.state.ny.us/03sosaddress/sos2003text.html (accessed July 7, 2005).

9.   www.dep.state.pa.us/newsletter/default.asp?NewsletterArticleID=9466 (accessed August 1, 2005).

10. www.dsireusa.org/library/includes/incentivesearch.cfm?Incentive_Code=RI05R&Search=Type&type =Purchase&CurrentPageID=2 (accessed July 11, 2005).

11. http://votesolar.org/sf.html (accessed September 28, 2005).

12. www.dsireusa.org/index.cfm?&CurrentPageID=2 (accessed September 28, 2005).

13. www.seattle.gov/mayor/climate (accessed September 13, 2005).

14. www.cleanair-coolplanet.org/information/ne_governors.php (accessed November 10, 2005).

15. www.ef.org/westcoastclimate (accessed November 10, 2005).

16. www.westgov.org/wga/initiatives/cdeac/index.htm (accessed November 7, 2005).

17. www.rggi.org/about.htm (accessed November 10, 2005).

18. www.chicagoclimatex.com/index.html (accessed September 20, 2005).

19. http://energy.senate.gov/public/ (accessed November 11, 2005) and www.eei.org/industry_issues/electricity_policy/federal_legislation/summary_title_xiii.pdf (accessed November 11, 2005).

20. www.eei.org/industry_issues/electricity_policy/federal_legislation/summary_title_xiii.pdf (accessed November 11, 2005).

THE COSTS OF RENEWABLE ENERGY TECHNOLOGIES

One way to express the cost of a clean energy purchase is in terms of additional cost per student per year. Bob Burhenn, director of energy and utilities management at the Catholic University of America, translated the university's wind purchase to the equivalent of buying one soda per month per student. The overall costs of a renewable energy project on or near a campus will vary considerably, depending on the availability of the renewable energy resource at a given site, the size of the project contemplated, the technologies employed, labor costs, financing costs (including the availability of financial incentives), and so forth. However, some general cost categories will come into play, and every institution will need to consider them when moving forward with a project.

This chapter will help colleges and universities decide which renewable energy sources they might consider for an on-campus (or near-campus) project. There are many intangible considerations, but this chapter will focus on costs. An alternative to costing out the project is to hire a contractor on a performance contract basis. This choice will reduce much of the analytical burden on the college or university and transfer the performance risk to the contractor. But it will also likely end up costing more, and because of the proprietary nature of performance contracts, it may not allow the college or university to include interested students and faculty in the details of the project.

The main renewable energy sources that colleges and universities consider are solar photovoltaics, wind power, and biomass (including landfill methane). Some colleges also consider geothermal energy, mainly of the low-temperature variety for direct heating and cooling. Few consider the high-temperature variety for electric power generation because very few colleges and universities have access to high-temperature geothermal resources. Other renewable sources, such as municipal solid waste or hydropower, are not typically considered because of environmental concerns. In addition, very few campuses have access to rivers or streams that are appropriate for hydroelectric generation.

FIGURE 3-1: COST OF RENEWABLE ENERGY PROJECTS

Energy Source	Cost (¢ per kWh)
Solar (cloudy climate)	64
Solar (sunny climate)	29
Biomass	4.5–8.0
Wind	3.8– 7.0
Landfill gas	3.7– 6.0
Municipal solid waste	3.5–4.5
Hydropower	3.2–3.5

Source: Compiled from U.S. Department of Energy, renewable energy trade associations, National Renewable Energy Laboratory, and data from individual projects

Solar photovoltaics (PV) are the most expensive renewable energy source (fig. 3-1). They are also the most common renewable energy technologies on campuses. Some campuses have also installed wind power and biofuel systems. The advantage of solar PV is that it allows for very small-scale installations and can be expanded easily by adding more PV modules. The disadvantage is that since solar PV is the most expensive form of renewable energy, if the college or university has a constrained budget, it will get less bang for its buck.

▶ STANDBY CHARGES

Almost all renewable energy projects face the possibility of standby charges. A standby charge or standby tariff is a fee imposed by the local electric utility to cover the cost of having extra electricity generation capacity ready in the event that the project (whether a renewable or nonrenewable energy project) breaks down or stops operating for any number of reasons. It is essentially an insurance charge the college or university pays to ensure that the electricity will be there if and when the project fails. If the campus is paying a standby charge and the renewable energy generator goes down, it will not face an increased demand charge.

Standby charges can be stiff. When Pierce College's local utility (Los Angeles Department of Water and Power) found out about the college's on-site PV-CHP (combined heat and power) generation plans, it added an annual standby charge of $50,000 to the electric bill. The utility added another charge of $5,000 per year because the college meets up to 10 percent of its power demand through self-generation.

Standby charges are controversial. They can not only be high, but they can also dramatically increase, eliminating the economic viability of a renewable energy project. Utilities often impose standby charges even when they do not own their own generation facilities and therefore are not generating the extra capacity necessary if the college or university's generator goes down. The controversy has led some states to regulate standby charges. Some regulators acknowledge that the intent of state energy policy is to support the development of renewable energy and standby charges should not create a barrier to such development. In certain states, utility regulators have ruled to exempt renewables from standby charges if they are under a certain size. Several states, including Rhode Island, Massachusetts, New York, California, and Connecticut, have completely exempt "clean" on-site generators from standby rates.1 In other cases, state regulations allow renewables to pay a lower standby charge than nonrenewable projects.

One way to eliminate standby charges is to sell the output of the renewable energy project directly to the utility. Under the Public Utilities Regulatory Act (PURPA), utilities are required to purchase electricity from "qualifying facilities," which include most renewable energy projects. In states that have net metering (see chapter 2), the utility is required to buy the renewable electricity at the same price as it is selling conventional electricity. This requirement forces a college or university to sell its renewable electricity to the utility and then buy it back at the same price, just to avoid standby charges.

Wind Power

As a rule of thumb, wind projects cost $1 million per installed megawatt ($1,000 per kW). This estimate will vary according to the wind regime in a given location and the availability of turbines (in 2004–2005, larger-than-expected demand and constrained supplies kept turbine prices high).

Price will also vary with the size of the turbines, which can be anywhere from a few kilowatts to 3 megawatts in capacity. Smaller turbines will cost far more per installed kilowatt than the large utility-grade turbines.

In deciding what capacity of turbine(s) to install, colleges and universities should consider net metering laws in their state. Nearly all net metering laws place limits on the capacity of renewable energy projects that can be net metered. If the wind project is within the limit, the college or university can sell electricity to the utility at the same price it is buying the electricity. (Net metering limits are rarely a concern for solar energy projects, which tend to have capacities well below the limits.) The net metering limits may not be a concern in instances where utilities are interesting in buying the wind power from the college or university, for example in states whose Resource Portfolio Standards require them to buy renewable energy. The price the utility offers will not be as high as the net metered price, but it may be high enough to make the wind project financially viable.

Up-Front Costs

A highly simplified breakdown of costs for a utility-grade 1.65 MW wind turbine is as follows:

$1.50 million	Wind turbine
.35 million	EPC contract
.01 million	Utility interconnection
.14 million	All other (pre-construction studies, land lease payments, etc.)
2.00 million	TOTAL

This is very approximate, as the costs of each component can vary considerably. The following are the main items to cost out in an economic analysis of a wind project:

Studies. A number of studies must be conducted prior to permitting a wind turbine. These include:

- Wind Study—Wind studies with anemometers on towers can run $25,000 to $75,000.

- Interconnection Study—This engineering study will cost approximately $5,000 to $50,000, depending on the size of the project, and can take 3–9 months to complete.

- Environmental Impact Assessment

- Soil Test—This test is needed to make sure there are acceptable subsoil conditions to accommodate the anchoring of the tower and determine the foundation design. The test should cost several thousand dollars.

- Telecommunications Impact Assessment—Because wind turbines can potentially interfere with radio, television, and microwave transmissions, they must be sited so that their towers and rotors are outside an exclusion zone that surrounds the microwave link path and antennae.[2]

- Historic Preservation/Cultural/Archeological Survey

- Permitting Appraisal—This determines what kinds of permits will be needed and their costs. Commonly required permits include building permits and county conditional/special use permits.

Turbine and tower. The main cost will be the wind turbine and tower. This cost will normally include installation, supervision, and commissioning. For a 1.5 MW utility grade turbine, the cost will run roughly $1.1 million to $1.6 million.

Balance of plant. This includes transformer, controls, cabling, and other miscellaneous equipment, which will run $300,000 to $500,000.

Interconnection with the utility's distribution system. The cost of the physical interconnection with the utility will depend on how close the project is to the utility's distribution lines.

Land acquisition or land lease. Negotiated lease payments to landowners for wind turbine towers are currently running roughly $2,000 to $5,000 per tower per year (or the payment can be a fixed percentage of gross revenues, which is the approach Carlton College used). The college or university will need approximately two acres for a 1.5 MW utility-grade wind turbine. This size will allow the turbine blades to avoid overhanging onto the adjacent property. During construction, additional space will be needed for construction equipment, plus there may need to be some space available to lay down the blades if they ever need to be removed from the turbine. It is standard practice to compensate the landowner for all crop and soil compaction losses after construction or extraordinary maintenance. The college or university will need to have a side agreement to cover this occasional trespass and compensation.

CASE STUDY

Carleton College—Earning Income with a Utility-Grade Wind Turbine

Carleton College's 1.65 MW wind turbine is making money for the college. The turbine is about a mile from the Northfield, Minnesota, campus, and all the power output (5.2 million kWh per year) is sold to the local utility for 3.3 cents per kWh. In addition, under a state incentive program, the college receives 1.5 cents for every kilowatt-hour produced. The college buys power from the utility at 5.3 cents. The output of the turbine is equivalent to 40 percent of the college's electricity consumption.

The project was initiated by a nonprofit organization and proposed to a campus environmental advisory committee of faculty, the college dean, and other staff. The board of trustees approved the project since it would at least break even. The college hired an attorney and alumnus who managed the process, along with an experienced contractor from South Dakota.

The project involved five contracts:
- Construction contract—noncompetitive
- Turbine procurement (includes maintenance)—competitive bid (3 suppliers submitted bids)
- Power purchase agreement with utility—20 years @ 3.6 cents
- Utility interconnection agreement
- Interconnection contract—a wind farmer who simply liked the idea of the college selling wind power to the utility provided free labor to connect the turbine to the utility grid.

Project Costs:
- Capital cost—$1.8 million
- O&M—$15,000/year—turbine manufacturer provides it
- Electricity purchase to provide power to sensors and computers when wind isn't blowing—$1,000
- Lease payment to landowner—2 percent of gross revenues
- Insurance—$12–15,000/year includes property insurance and business interruption insurance because the college has a contract with the utility to provide power 80 percent of the time

Access road. In most cases, the college or university will need to construct a gravel access road to the wind tower. This will run approximately $10 to $15 per foot.

Warranty. A two-year warranty on parts will run approximately $20,000.

Annual Costs

Operations and maintenance (O&M). This includes replacement equipment/parts, testing, and down time. The cost of down time translates into either a loss of electricity sales to the utility if the electricity is being exported, or into increased retail electricity purchases from the utility if the electricity is consumed on campus. A vendor can provide an O&M package for about $15,000 to $20,000.

Extended warranty. A five-year extended warranty—that is, a warranty that runs from the end of the 1–2 year manufacturer's warranty on up through year five—will cost $25,000–$30,000 per year.

Insurance. Insurance will run $15,000–$18,000 per year during the warranty period. Premiums will increase after the warranty period and will continue increasing into the future as the likelihood of problems and breakdowns increase. During the warranty period, insurance will cover damage from lightning strikes and power quality problems arising from the interconnection to the power grid. After the warranty period, the insurance will cover liability, business interruption, and operational risks. There will be a deductible, typically approximately $20,000 per event.

Land lease. (See above under "Land acquisition or land lease.")

Other annual costs. Other costs might include access road maintenance, meter reading, permit renewals or other government fees, and finance charges.

▶ SOLAR PHOTOVOLTAICS (PV)

Solar PV projects are the most common form of renewable energy installed on college campuses, in part because they can be small and relatively inexpensive. They are also modular and can be easily expanded, and they do not involve combustion or moving parts. Many colleges have small, 1–15-kW PV systems for educational or demonstration purposes. A number of colleges are installing 200-kW PV systems and larger. The size of the system a college chooses depends on several factors: affordability, especially after factoring in government grants and rebates; state limits on net metering; and the power demand of the facility where the PV system is being installed. For example, Arizona State University selected a PV system of 30 kW to power daytime lighting in a campus parking garage in order to match the electricity demand for lighting, which is approximately 30 kW.

    As a rule of thumb, solar photovoltaic (PV) installations cost approximately $5.50 per watt ($5,500 per kW). For example, a 250-kW roof-mounted system would cost $1.375 million. Transaction costs will add about 3 percent to the cost, as shown in figure 3-2.

FIGURE 3-2: ESTIMATED TRANSACTION COSTS FOR SOLAR PV PROJECTS

	250 kW roof-mounted system	1 MW ground-based covered parking lot system	1 MW open field system
Environmental assessment	$5,000	$7,500	$8,500
Project management	$15,000	$20,000	$15,000
Design review and approval	$5,000	$7,500	$6,600
Due diligence	$7,500	$10,000	$10,000
Construction inspection	$4,000	$4,000	$4,000
Total	$36,500	$49,000	$44,100

Source: California State Consumer Power and Conservation Financing Authority, State Facility Solar Power Purchase Program, RFP of April 22, 2004, p. 14.

CASE STUDY

Pierce College—Solar PV System Installed on a Performance Contract Basis

Pierce College's $2 million, 191-kW solar photovoltaic system was packaged with a 360-kW microturbine cogeneration system for a total cost of $4.98 million. The California Public Utility Commission provided an incentive payment of $2 million through the local electric and gas utilities. In addition, the California Energy Commission provided a $1.3 million low-interest loan, and the college put up $350,000 in cash. The balance was financed on a 15-year performance contract with Chevron Energy Solutions. The project was estimated to reduce the college's electric bill by 30 percent. The monthly performance contract payments started out at $270 per month, and the estimated utility bill savings are $250 per month. Shortly after the loan was established, voters approved a bond issue, and Pierce used the revenue to pay off the balance of the performance contract and the California Energy Commission loan.

Pierce was motivated to examine renewable energy options in 2002 when the Los Angeles Community College District (LACCD) Board of Trustees adopted a policy requiring that all new projects be powered using on-site renewable energy.[3] Before deciding on the solar system, Pierce examined a variety of other technologies. The college settled on PV because of the availability of the state government subsidy. The decision was also based on projections that the solar system would reduce both energy costs (especially midday peak demand costs) and the generation of greenhouse gases while also providing an extra backup power supply for periods when the grid is incapacitated. In fact, the PV-CHP system is sized to provide all the campus's backup power except for air conditioning. The heat from the CHP units fuels the chillers for air conditioning, providing 100 tons of the 800-ton system. The PV system was installed over a parking lot, providing the added benefit of shading cars that park underneath the panels.

Chevron Energy Solutions was selected as the project manager and performance contractor through a competitive bidding process. Chevron made the decisions on all procured equipment, including the PV arrays. The PV-CHP system provides power directly to the campus grid, and no power is sold to the local electric utility, the Los Angeles Department of Water and Power, which charges Pierce College a standby charge for self-generation (see page 28).

Upfront Costs

- Design costs
- Solar modules
- Support structures, inverters, transformers, controls, and other equipment
- Batteries (optional)
- Maintenance contract (unless maintenance paid annually out of maintenance budget)

Annual Costs

- O&M (unless maintenance contract paid upfront)
- Batteries (optional)—replaced roughly every 10 years
- Insurance

PV can be incorporated into the design of new campus buildings. In cases where PV systems are deemed too expensive, design elements can be incorporated to make the building "PV-ready." For example, the University of British Columbia's C. K. Choi Building, designed in 1994, does not have a PV system but is oriented toward the sun, has a reinforced roof, and has special accommodations for housing system equipment (e.g., inverters, transformers and controls) and a battery bank.[4]

Pros and Cons of Batteries

Most colleges and universities do not install battery banks or other electricity storage systems with their renewable energy projects. Battery banks add a considerable cost to a project. The main advantage is round-the-clock backup power for critical functions in the event of a grid outage. Battery systems are usually associated with solar PV systems. They provide power when the sun is not shining and are recharged by the PV system when the sun is shining.

The majority of colleges and universities install diesel or gasoline generators for backup purposes. Such units have far lower first costs than PV systems, particularly if the PV system has a battery bank. But during extended grid outages, the conventional backup generators can run out of fuel, and depending on the type of outage or emergency, fuel deliveries may not be possible.[5] In addition, conventional backup generators operate only during outages and do not pay for themselves, while a PV backup system, which does not require fuel purchases, can be operated continuously and pays for itself over time through reduced grid power purchases. Continuous operation allows the facilities staff to fix problems as they emerge, while problems with a conventional backup generator only show up during an emergency situation.

A battery system can add as much as 30 percent to the front-end cost of a PV system. In addition, the batteries will have to be replaced at regular intervals, roughly every 10 years assuming proper maintenance. The PV system should be sized so that it can provide electricity to the critical end users while also charging the battery bank. It should be large enough so that it can perform both tasks on days with no sun.

▶ BIOMASS

Biomass is perhaps the most complex of the renewable energy sources for a college or university to investigate because it encompasses a range of fuel types fed into a range of conversion and power generation technologies. The fuel types include crops and crop waste, animal and human waste, wood and forest residues, and methane gas from landfills and sewage treatment plants. These fuels can be used directly or converted—for example, from a solid to a gas—to use in certain power generation technologies, such as gas turbines, fuel cells, and Stirling engines. They can also be cofired alongside conventional fuels in existing power plant boilers. Cofiring may be the most cost-effective way to use biomass because the power generation infrastructure is already in place and the only expenses are biomass feedstock, its transport, and a system to store and inject the biomass into the boiler.

Biomass fuel can originate in the form of solids (e.g., wood chips, agricultural waste, and urban wood waste), liquids (e.g., ethanol from crops or agricultural waste), or gases (e.g., methane from sewage treatment or landfills). All three forms can be used as feed-stock for various power generation technologies. Solid fuel is often the least expensive since it requires little processing prior to combustion. Most of the 10 gigawatts of installed biomass power generation in the United States use direct combustion of solid biomass fuels. However, solid fuels are the least efficient and dirtiest form of biomass. They can leave more residues in the combustion process, which translates into a higher maintenance expense. Liquid forms of biomass can replace all or a portion of the gasoline or diesel fuel used in generators. Using gasified biomass for electricity production is up to twice as efficient as burning it in solid form and results in lower carbon emissions. Gaseous biomass fuels are usually the most efficient, but if the starting point is solid biomass fuel, then the gasification process can be expensive. If, on the other hand, the starting point is gas, such as methane from animal waste or sewage, the costly gasification step is unnecessary. Instead, a less costly technology to capture the gas, such as an anaerobic digestion system, will be needed. Biomass must be in a gaseous form to power the hydrogen production in a fuel cell.

The three basic biomass cost categories are capital costs, fuel costs, and operations and maintenance (O&M) costs. Other costs, such as taxes and insurance, are generally embedded in the O&M costs. The fuel costs incorporate the cost of transporting the biomass fuel to the power plant. In some cases, there can be additional costs such as fuel inventory and pollution control.

Biomass Fuels

Unlike other forms of renewable energy like wind and solar energy, biomass projects involve an ongoing fuel cost. Biomass fuels can be cost-competitive with conventional fuels (figure 3-3), but the cost is also highly dependent on the proximity of the fuel to

FIGURE 3-3: SAMPLE COST COMPARISON OF FUEL TYPES

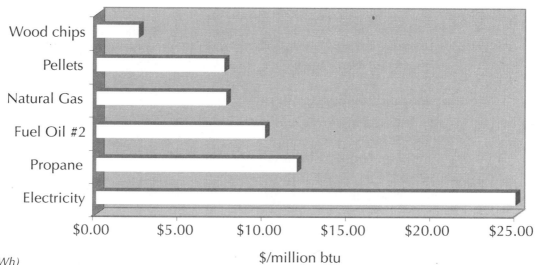

(3,143 Btu = 1 kWh)

Source: "Wood Biomass for Energy," Techline, Forest Products Laboratory, U.S. Forest Service, Madison, Wisconsin, 2004

the project. Transportation costs can undermine the economics of a project. A rough rule of thumb is that the biomass resources should not have to be transported more than 50 miles. With higher gasoline prices, that distance shrinks.

The average fuel price for biomass feedstock in U.S. biomass plants is 2.5 cents per kWh. This price includes the cost of transporting the fuel. Some plants get some of their fuel for free (not including transportation), but not all of it. And in a deregulated, competitive market, there is less frequency of free fuel. In some areas, such as Maine, biomass plants compete against each other for feedstock fuel. As a result, data on a biomass plant's fuel costs are considered proprietary and are increasingly difficult to obtain.

In general, biomass fuel from mill wastes and urban wood wastes ranges from $0 per MBtu to $1.40 per MBtu; agricultural waste is about $1 per MBtu; and forest residues are around $2.40–$3.50/MBtu.

Biomass fuel costs can vary over time. Unless a college or university is supplying all its own biomass fuel, such as Morrisville State College of the State University of New York with manure from its dairy barn, it will face biomass price fluctuations in the market-place. It is possible to mitigate the risk of shifting fuel prices by entering into long-term fixed-price contracts for a supply of biomass material, such as woodchips. But while such contracts provide price predictability, being locked into a long-term price can also mean the college or university will not be able to take advantage of lower-priced supply that may become available on a temporary or "spot" basis. For example, if someone is tearing out an orchard, it might present a one-time opportunity to buy biomass fuel—in this case, tree waste—at a very low price. The most economically successful U.S. biomass power projects are those that have a combination of long-term (one- to three-year), medium-term (one-year), and short-term (spot market) fuel supply contracts. Some commercial biomass plants operate solely with medium- and short-term contracts. Being able to buy fuel from a variety of sources means that, if possible, the biomass plant should be designed and licensed to accommodate a broad range of biomass feedstock fuels, from forest residues to agricultural waste to urban wood waste. In short, there is an art to managing biomass fuel supply contracts and purchases.

Wood, Agricultural Crops, and Crop Waste

Wood chips and forest waste are common feedstock fuels for biomass plants. They are particularly common in central heating and cooling plants, as opposed to power-generating plants. Many of the heating plants at K–12 schools in Vermont are fueled with wood chips, while Chadron State College in Nebraska burns sawmill waste and residual matter from commercial timber harvesting operations. Central Michigan University is using wood chips in its central heating and cooling plant, thereby replacing natural gas and saving over $1 million a year in energy costs. The university has a two-year fixed-price contract with a wood supplier, although the price can be (and has been) adjusted for increased transport fuel costs.[6] Note, however, that forest waste is not an eligible resource for the Center for Resource Solutions' (CRS) Green-e certification, meaning that RECs might not be as tradable or sellable. However, CRS will defer to local environmental groups in many cases.[7]

CASE STUDY

University of Iowa—Cofiring Biomass with Coal

Cofiring biomass with conventional energy sources such as coal is perhaps the most cost-effective way to use biomass for power generation. In the case of the University of Iowa, which has its own coal-fired power plant, cofiring is creating substantial dollar savings. The university is saving more than $500,000 in fuel and associated costs. In 2003, the university entered into an agreement with Quaker Oats to purchase oat hulls, a byproduct of the breakfast cereal manufacturing process. The oat hulls are delivered to the university, injected into its fluidized bed combustion turbine, and burned with the coal to generate electricity. The oat hulls have turned out to be far less expensive than the university's other energy sources (fig. 3-4). In addition, the oat hull price is not subject to the volatility of fossil fuel and electricity prices, thus providing the university with a hedge against price shocks.

To accommodate the oat hulls, the facilities staff designed a system for injecting the hulls into the coal boiler. Partly because of the injection system design, the plant can burn a 50:50 mixture of biomass to coal (as measured by heat content), while most cofiring coal plants are only burning biomass at 5–10 percent of the mix.

The plant has earned two Environmental Excellence Awards from the governor of Iowa.

FIGURE 3-4

Cost of Fuels Used by University of Iowa

Energy Source	Unit Cost ($/mmBtu) FY 04–05	Unit Cost ($/mmBtu) FY 05–06 (est.)
Biomass	$1.25	$1.25
Fluid bed coal	$1.83	$2.25
Stoker coal	$2.83	$3.85
Natural gas	$7.44	$9.33
Electricity from utility	$10.96	$11.03

Source: University of Iowa

Landfill Methane

Capturing and burning the methane that is naturally produced at solid waste landfills is among the most economical ways to generate electricity with renewable energy. Many electric utilities and municipalities have tapped landfills for this purpose. Since colleges and universities do not own landfills, using landfill methane involves entering into an agreement with either the local government that owns the landfill site or a private firm that has an agreement with the local government to develop the site for methane recovery.

UCLA has been burning landfill methane on campus for more than 10 years. UCLA entered into an agreement with GSF Energy, Inc., operators of a landfill gas recovery plant, to purchase methane gas and have it piped 4.5 miles to campus. There, the gas is compressed, blended with natural gas, and burned in a 40-MW cogeneration facility.

The facility would have run solely on natural gas, but now that methane is also used, UCLA saves $250,000 a year on fuel purchases. The savings are increasing as natural gas prices rise.[8]

It was decided that the best course of action was to isolate the project's funding from external factors by issuing Certificates of Participation to lenders. The certificates are essentially loan agreements paid back with the operating savings realized by the new system. This approach allowed UCLA to ensure that the cost of borrowed funds accurately reflected the financial soundness of the project. The cogeneration system required an initial investment of $188 million, which will be paid off over 22 years. After that, it will provide savings of more than $25 million over its anticipated life span.

Anaerobic Digestion Methane

Organic human or animal waste can be put into an anaerobic digester, which captures the methane from the breakdown of the waste. The methane can then be burned for

CASE STUDY

Hudson Valley Community College—Landfill Methane Piped to Campus Generator

In 1996, the city of Troy, New York, was planning to cap a landfill that was contiguous with the campus. Hudson Valley Community College examined the possibility of using the landfill methane for the campus boilers, which provide space heating for buildings. The examination found that the cost of piping the methane from the landfill to the various boilers around campus, plus the cost of filtering the gas, would be too high. Next, the college examined the possibility of using the methane for power generation. A consulting firm was hired and a study found that the power generation could be done economically.

The college installed four generation units—three 1,350-kW natural gas-fired units and one 800-kW methane-fired unit, with a pipeline from the landfill to the generator. These units supply all the electricity needs of the campus, which is now completely disconnected from the utility grid. The methane-fired unit supplies 33 percent of the campus's electricity demand in the winter and 25 percent in the summer, when air conditioning costs drive up demand and the natural gas units' output must be increased. Waste heat is used to power an air conditioning chiller for the gymnasium. It is also used to heat the gym in the winter. The cost to heat and cool the gym was $90,000 a year. Now it is zero. The college is considering expanding the heating loop to other buildings on campus.

The overall capital cost was $8.4 million for four generation units, a 2.2-MW diesel backup unit, the landfill pipeline, the building housing the generation units, the transformers and controls, the waste heat loop for the gym, and a 15-year, $825,000 performance contract with Siemens that guarantees the college a "profit" of $1.5 million over the term of the contract.[9]

Hudson Valley Community College received a grant from the New York State Energy Research and Development Authority and is expecting a $2 million grant from the state Department of Environmental Quality under a program supporting landfill closure and methane gas management. The remaining money was financed through a 15-year performance contract with Siemens. The debt service on the performance contract is paid out of the savings on the electricity bills. The college would have saved if natural gas prices had remained flat. But the spike in gas prices has wiped out most of the savings. Of course, once payments under the performance contract end, the college will capture all of the savings.

space heating and cooling purposes or to generate electricity. The capital cost of anaerobic digestion varies greatly across operations from $1.40 to $8 per watt depending on the number and type of animals used.[10]

To determine the cost of an anaerobic digester combined with an average gas-fired power generator, some rules of thumb can be applied. A chicken produces about .90 pounds of manure per day; a human produces about 2.5 pounds; a hog produces about 9 pounds; a feedlot cow produces about 90 pounds; and a dairy cow produces about 140 pounds.

Approximately .90 pounds of manure yields one watt of power for a day. Therefore, 900 pounds from any source yields 1 kW of power generation per day, or 24 kWh. As such, one chicken yields .90 watts; one human yields about 3 watts; one hog yields 10 watts; one feedlot cow yields 102 watts; and one dairy cow yields 155 watts. Thus, 5,000 chickens can produce 5 kW; 5,000 hogs can produce 50 kW; a 5,000-head feedlot operation can produce 512 kW; a 5,000-head dairy farm can produce 775 kW; and 5,000 people can produce 150 kW.

An anaerobic digester (with small-scale single-cell lagoon and 75-kW generator) will cost roughly $375,000 and will reduce electricity bills by about $40,000 per year (assuming an electricity cost of 6.75 cents per kWh). This calculation assumes no state or federal grants or other subsidies.[11]

Given the large number of animals (or humans) needed to produce a small amount of electricity, it may be worthwhile to examine the possibility of a community methane digester. With increasing state regulations on odor control, farmers are looking for solutions to their animal waste problems, and methane digestion could provide such a solution.

Morrisville State College of the State University of New York is using manure from its dairy farm as fuel for anaerobic digestion. The system handles manure from 250 milking cows, which is on the lower end of what is economically feasible. The milking cow waste is supplemented by additional waste from the heifer and calf barns, and the college is examining adding organic waste from the main dining hall. The manure is fed into the digester, and the captured methane is used to fuel a 55-kW Stirling engine for electricity generation.

SUNY Morrisville had considered capturing the methane from the campus wastewater treatment plant, but the modifications necessary to convert the plant to anaerobic digestion were too expensive for the amount of methane that would have been generated. Most campuses have so far rejected converting to anaerobic sewage treatment because it is too costly unless subsidies are available. The process requires a large supply of organic matter to make the economics work. According to one EPA analyst, it takes a city with a population of about 90,000 people to make a sewage-based anaerobic digestion system pay for itself through reduced energy consumption in the sewage treatment process and the generation of electricity for sale at average prices.[12]

Biomass Combustion Technologies

Biomass fuels can be used to generate electricity or simply to fuel a boiler to provide heat for campus buildings. When used for power generation, biomass fuels work with some of the same power production technologies as fossil fuels. Most biomass plants use steam turbine technology. Other generation technologies are in use, but mainly on a noncommercial demonstration basis.

FIGURE 3-5: COSTS OF POWER PRODUCTION TECHNOLOGIES

Technology Type	Cost/kW	Annual O&M Cost per MWh	Size Range
Diesel turbine	$500–600	$5–10	1kW– 0MW
Natural gas turbine	$600–1,200	$7–15	1kW–250MW
Steam turbine	$800	$4	50kW–300MW
Combustion turbine	$650–900	$3–8	300 kW–25 MW
Microturbine	$300–500	$2–10	25kW–500kW
Stirling engine	$700–1,100	$2–8	<1kW–25kW
Fuel cell	$3,000–6,000	$3–15	1kW–250kW

Sources: U.S. Department of Energy, The Energy-Smart Guide to Campus Cost Savings (Washington, DC: 2003), www.rebuild. org/attachments/solutioncenter; and the Consumer Energy Council of America

Fuel Cells

Fuel cells are not a form of renewable energy, but they can be fueled by renewable energy such as biogas. They are an energy-efficient way to generate electricity, using about half the fuel of conventional power plants per kilowatt-hour generated. They are also clean, with the only emission being water vapor. However, fuel cells have a high up-front cost, and most purchasers need a significant subsidy.

Most fuel cells run on natural gas, although it is possible to run them on biogas derived from wood, crops, sewage treatment, or other bioenergy resources. A few colleges and universities have installed fuel cells and generally run them on natural gas. The one notable exception is the State University of New York College of Environmental Science and Forestry, which is planning to run its fuel cell, at least in part, on biogas.

Operations and Maintenance

O&M costs generally are both fixed and variable. They vary with the operational status of the plant. O&M may increase when more kW are produced due to increased use of equipment. Therefore, the cost of O&M per kWh essentially depends on how many kilowatt-hours are generated. Primary energy consumption and costs may also increase with increased production. Government incentives can help drive down the cost of a plant's electricity, increasing sales and thus lowering the cost of O&M costs per kWh.

O&M costs per kWh are actually a crude indicator of plant economics. Ultimately, the structure of the electricity sales contracts will determine the economics of a biomass plant.

CASE STUDY

SUNY College of Environmental Science and Forestry—Biogas-Fueled Fuel Cell

In late 2005, the State University of New York College of Environmental Science and Forestry (SUNY-ESF) installed a $1.35 million 250-kW fuel cell to generate on-site electricity and provide residual heat for domestic hot water and heating at its campus in Syracuse. The fuel cell, which will provide 6–8 percent of the campus's power, is currently fueled with natural gas, but plans are underway to use fast-growing willow trees as a source of feedstock fuel in the future. A gasifier will be installed to gasify the wood from willow trees. Initial calculations indicate that the biogas might be competitive with natural gas. The percentage of biogas in the fuel mix will start low and be gradually increased based on evaluations of the fuel cell's performance.

In addition to providing electricity, the fuel cell will produce heat for the building in which it is housed. The combined heat and power efficiency of the unit is 70–80 percent, which means that 70–80 percent of the energy contained in the feedstock fuel is converted to useful energy, compared to 33 percent for grid-supplied electricity.

The university did not pursue the fuel cell project simply for on-site generation purposes. Rather, it considers the project to be a component of its research and development work. The project is not intended to demonstrate the economic feasibility of fuel cells or biomass gasification. Thus, the university is financing the project almost entirely with research grants. To cover the $3.5 million cost, it received a $1 million grant from the New York State Energy Research and Development Authority (NYSERDA), $250,000 from the U.S. Department of Energy's fuel cell program, and $75,000 from the Electric Power Research Institute, the research arm of the nation's electric utility industry. The remaining cost is financed through a loan from the New York Power Authority. The college's capital fund will be used to cover the debt service on the loan and residual capital value. FuelCell Energy of Danbury, Connecticut, supplied and is maintaining the fuel cell.

Total Costs

The total cost of owning and operating a biomass power plant in the United States averages about 6.5 cents per kWh, assuming an 80–100 percent capacity factor. This breaks down into almost equal fuel, capital, and O&M component costs. With not much variation among operating plants with respect to O&M costs or cost per installed kilowatt, the major factor affecting total cost is the availability of government subsidies and/or capacity payments from the local distribution utility to keep the plant ready to run. There is also some variability in fuel costs. Urban wood waste tends to be the cheapest fuel, although proximity to sawmills can drive fuel costs to less than 1 cent per kWh.

In some cases, the waste ash from the plant can be a source of revenue. The ash marketing and sales of a commercial biomass plant in Tacoma, Washington, for example, saved the plant $600,000 per year compared to the cost of ash disposal.

Fuel flexibility is important in keeping fuel costs down. It is also important that the plant be designed and permitted from the start to accept a range of feedstock fuels. Many existing commercial biomass plants that currently accept a narrow range of fuels are looking at expanding the types of fuels they will burn.

It is advisable to have some short-term (one-year) contracts or even some spot market purchases. Fuel costs vary significantly, and while long-term multiyear contracts provide cost predictability and comfort to college and university business officers, they close off the opportunity of lower-cost fuel that a spot market provides.

Some fuel supply and other risks can be mitigated through joint ownership of the biomass plant. More than most other renewable energy technologies, biomass projects lend themselves to cooperative efforts with members of the community, such as farmers and ranchers, who can contribute agricultural waste, dairy waste, and other fuels and can co-finance and co-own the power plant.

▶ ## GEOTHERMAL ENERGY

Geothermal energy uses heat from the earth to generate electricity or provide heating and cooling for buildings. Wells are sunk into the ground to bring steam or hot water to the surface, where it is run through equipment that turns it into useful energy. There are two basic types of geothermal energy: high-temperature and low-temperature. High-temperature geothermal energy uses the geothermal steam to turn an electricity-producing turbine. This method is only economical in places where the geothermal resource is relatively close to the surface. The main high-temperature geothermal resources in the United States are found in Nevada, California, and Hawaii.

Low-temperature geothermal resources are found throughout the country and are being used in 23 states. They are not hot enough to turn a turbine, but they can be used in geothermal "heat pumps" for space heating and cooling, thereby offsetting the campus's electricity charges for these applications.

The Oregon Institute of Technology has been using low-temperature geothermal energy since 1964. Three hot water wells drilled during the original campus construction vary from 1,300 to 1,800 feet. These wells supply all of the heating needs of the 11-building, 600,000-square-foot campus as well as a portion of the cooling needs. OIT's geothermal system costs $35,000 to operate each year, including maintenance salary, equipment replacement, and cost of pumping.[13] This cost compares favorably to the operating cost of a natural gas-fired boiler at $250,000 to $300,000 per year.[14] Oberlin College incorporated a low-temperature geothermal system into the design of its Lewis Center for Environmental Studies. The 24 geothermal wells, at a cost of $90,000 each (for a total of $2.16 million) supply 100 percent of the building's heating and cooling needs.

Because low-temperature geothermal energy does not generate electricity, utilities will get no credit under state renewable portfolio standards (RPS) for supporting it. But because low-temperature geothermal heat pumps serve to reduce demand for grid power, they are energy-efficiency measures. By late 2005, several states—Colorado, Connecticut, Hawaii, Nevada, and Pennsylvania—had incorporated or were in the process of incorporating energy-efficiency measures into their RPS rules. Colleges and universities interested in getting financial support from utilities for low-temperature geothermal applications should check with their state energy offices or public utility regulatory bodies to see if such applications qualify under the RPS.

ENDNOTES

1. Peregrine Energy Group for the Clean Energy Group "Standby Charges and Fuel Cells: New Opportunities for State Policy Coordination," September 2005, p. 1.

2. Consult International Telecommunications Union, "Assessment of impairment caused to television reception by a wind turbine," Recommendation ITU-R BT.805, 1992.

3. The District of Columbia also adopted a policy to ensure that all of its new buildings be LEED-certified.

4. Jennifer Sanguinetti, Keen Engineering, personal communication, August 2005.

5. For example, during the 2004 hurricanes in Florida and during Hurricane Katrina in Louisiana in 2005, backup diesel generators ran out of fuel.

6. George Ross, vice president for finance and administrative services, Central Michigan University, personal communication, December 2005.

7. Scott Haas, McNeil Technologies, "Community Bio-Energy Applications: Providing a Market Outlet for Forest Thinnings," presentation at *Colorado Wind & Distributed Energy: Renewables for Rural Prosperity* conference, April 13–14, 2004. available at: www.state.co.us/oemc/events/cwade/2004/presentations.htm

8. www.districtenergy.org/CHP_Case_Studies/ucla.pdf

9. The project has increased the college's reliance on natural gas, and some of its "profit" has been eroded by the increase in natural gas prices.

10. Ed Lewis, Colorado Governor's Office of Energy Management and Conservation, presentation at Anaerobic Digestion Workshop, October 2004, available at: www.state.co.us/oemc/events/anaerobic/index.html

11. Ibid.

12. Patrick Kelly, Air Quality Analysis Section, U.S. Environmental Protection Agency, Region 6, Dallas, personal communication, August 2005.

13. Kevin Rafferty and Paul J. Lienau. "Geothermal Heating System, Oregon Institute of Technology Introduction" (Klamath Falls, OR: Geo-Heat Center, 2005), available at: http://geoheat.oit.edu/public/tp26.htm.

14. Renewable Energy Policy Project. "Geothermal FAQ," available at: www.crest.org/articles/static/1/995653330_5.html#geoh.

FINANCING A RENEWABLE ENERGY PROJECT

A range of grants, rebates, and financing approaches are available for on-site renewable energy projects. There are also some creative ways to pay for the renewable energy certificates (RECs) that are available from electric utilities and green power marketers and brokers. The first part of this chapter is a compendium of internal financial resources, such as large gifts and student fees, and external resources, such as private equity and lease products. The second part of the chapter discusses ways of minimizing the cost of a green power purchase.

▶ COMBINING SEVERAL SOURCES OF FINANCING

Unless an on-campus renewable energy project is very small (e.g., a few kilowatts of installed capacity) or is strictly a research project, it is unlikely to receive 100 percent grant financing. Most medium- and large-scale campus projects are funded from a combination of sources. Government grants are combined with funds from operating and/or capital budgets, and some debt may be involved. For larger campus wind projects, private equity investors with a tax appetite may be involved. Even when the up-front costs of a project are reduced through operating leases or performance contracts, typically there is still a combination of financial sources.

For example, a large wind project can often use a combination of state and federal grant sources to cover a third of the capital costs. Half or more of the costs can be covered with debt. The amount of debt and its terms can be structured so that revenue from the wind project's electricity sales and REC sales are sufficient to service the debt. The remaining capital cost can be covered with allocations from the capital budget, from a large gift, or, in this example, with an equity investment from investors attracted to the tax benefits that are unavailable to the institution.

▶ FEDERAL GRANTS, REBATES, AND INCENTIVES

Most federal government grants and tax incentives are technology-specific. There are different types of expenditures depending on whether a college is interested in solar energy, wind power, or other specific renewable energy technologies. Even within a single renewable energy source, there are different programs and subsidies. For example, most federal support for solar energy implementation is for solar photovoltaics, as opposed to other solar power generation technologies or solar water heating systems.

U.S. Department of Energy

The U.S. Department of Energy (DOE) supports renewable energy through a range of research & development (R&D), energy efficiency, and energy information programs. DOE's Office of Energy Efficiency and Renewable Energy (EERE) partners with universities, industry, and state and local governments. EERE made $506 million available in fiscal year 2004. The majority of the financial opportunities are for business, industry, and universities. These include grants, cooperative agreements, continuation awards, and renewal awards.[1] Most of the funds for renewable energy projects go to support research and development, not commercial or near-commercial applications.

DOE's e-Center provides a Web site for business and financial assistance opportunities at http://e-center.doe.gov/. EERE's Golden Field Office issues solicitations for funding opportunities posted at http://www.grants.gov/. For technical research projects, DOE's National Renewable Energy Laboratory (NREL) works with colleges and universities through its Technology Partnership Agreements. These are collaborative or "work-for-hire" type arrangements.[2] In addition, DOE's energy-efficiency programs can help campuses with energy management investments, which can free up funds for renewable energy. (See a discussion of savings from energy efficiency in this chapter.)

U.S. Department of Agriculture

The Renewable Energy Systems and Energy Efficiency Improvements Program—section 9006 of the Farm Bill of 2002—provides loans and grants to encourage agricultural producers and small rural businesses to create renewable and energy-efficient systems. In 2005, the program made available $200 million in guaranteed loan funds. In addition, competitive grants are available to cover up to 25 percent of the project cost for renewable energy systems. This program is for farmers, ranchers, and rural small businesses, and the owning party must be a rural small business. The renewable systems need to be located in rural areas. A college or university could partner with the farmers, ranchers, or rural small businesses and help with the grant application process, develop curriculum, and buy the renewable power.[3]

The Biobased Products and Bioenergy Program finances technologies needed to convert renewable farm and forestry resources into affordable electricity, fuel chemicals, pharmaceuticals, and other materials. Loans are eligible for financing under the Business and Industry Guaranteed Loan Program.

The Woody Biomass Grant Program in the U.S. Forest Service (a USDA agency) made $4 million available in 2006 for grants that increase the utilization of woody biomass from or near National Forest System lands. The grants are in the range of $50,000 to $250,000 each.[4]

Federal Appropriations

Many colleges and universities are familiar with appropriations earmarks that support research and capital projects. Some institutions, working through their congressional delegations or lobbying firms, have received federal earmarks for renewable energy projects. Iowa Lakes Community College received a $500,000 earmark for its 1.65-MW wind power project and another $500,000 for a laboratory and classroom for its renewable energy educational program. Saint Francis University in Pennsylvania received a $500,000 earmark for its wind power program involving a range of educational and investment activities.

There is no formula for how to win an earmark. Generally, the president speaks with members of the congressional delegation. It helps if a member is either in a leadership position in Congress or sits on House or Senate appropriations committees. Members with seniority also may have more influence. Although the majority party in Congress has more influence over appropriations, party affiliation does not matter as much for small earmarks. It also helps if a renewable energy project can be linked to a broader educational and/or research program at the college or university and if a portion of the earmark is proposed to support that program.

Wind Energy Production Tax Credit

The wind energy Production Tax Credit (PTC) provides a subsidy of 1.9 cents per kWh to owners of wind projects over the first 10 years of operations. It can only be used against corporate income, so government or nonprofit institutions such as colleges and universities cannot benefit directly from it. However, deals can be structured, particularly for larger projects, to attract private investors who are looking for tax shelters and who will provide equity investment in exchange for the tax benefits. (See a discussion of private equity investors in this chapter.)

To qualify for the PTC, a wind project must sell its electricity to an electric utility. This requirement does not rule out a college or university using its wind generation for its own electricity needs, but it adds an extra step. The institution will need to negotiate with its local utility to buy the wind turbine's output and then sell back an equivalent amount. The ideal situation would be for the utility to buy and sell the electricity at the same price. If the turbine's output (or installed capacity) is within the state's net metering limits, then the utility will be required to buy and sell at the same price (see "Net Metering," chapter 1). But this is highly unlikely to be the case for any turbine large enough to attract investors seeking the PTC tax shelter. That is, a wind turbine large enough to attract private equity investors (at least 1.5 MW) may exceed state limits on net metering.[5] This means that in negotiating the buying and selling prices with the utility, there will be no state law requiring the utility to buy the output at a retail price. However, the economics of the turbine may still prove to be positive even with the utility buying the output at a wholesale price and selling it back to the college at the retail price.

The wind energy production tax credit expires every few years, and Congress must act to renew it. In the past, the PTC has expired before Congress renews it, with the result that wind power development slows as developers and investors wait for Congress to act. To qualify for the PTC, a wind project must be up and operating prior to the PTC's expiration. To keep abreast of the status of the PTC, visit the American Wind Energy Association Web site at www.awea.org.

Solar Energy Tax Credits

States have a range of solar energy tax credits, and the 2005 Energy Policy Act provides a 30 percent federal solar tax credit that applies to the investment balance remaining after any state (or utility) incentives. The federal tax credit is available for projects installed by December 31, 2007. After that, it will expire unless Congress reauthorizes it.

As with the wind energy production tax credit, solar energy tax credits are not available to nontaxpaying institutions like colleges and universities. However, these institutions can attract private investors who are willing to own the solar installation in exchange for the tax benefits. (See a discussion of third-party service models in this chapter.) Colleges may also have alumni interested in supporting their alma mater while receiving tax benefits. Investors will be most attracted to solar investments in states with the most generous solar subsidies: California, Connecticut, Hawaii, Illinois, Massachusetts, Nevada, New Jersey, Oregon, and Rhode Island. New incentive programs in Texas, New Mexico, Arizona, Colorado, and Washington, DC, could make these states attractive as well.

▶ STATE GRANTS, REBATES, AND INCENTIVES

Thirty-seven states and the District of Columbia have financial incentive programs for renewable energy (figure 4-1). Like federal support, much state support is specific to the type of renewable energy technology being implemented. In many cases, state support will only cover a small percentage of project costs, and there will be a dollar limit on the amount given to any one project. Some incentives are only available in certain sectors, such as the residential or commercial/industrial sectors. Colleges and universities should check with their state energy offices and public utility commissions for eligibility requirements.

State Energy Funds

Fifteen states have energy funds or "public benefit funds" capitalized by surcharges on electricity sales within the state (figure 4-2). The funds support renewable energy and energy-efficiency projects and enterprises, among other actions. They are usually in the form of grants, but in a few states a portion of the support is provided in the form of loans. Because public benefit funds rely on a surcharge and not on annual appropriations, they tend to be less restrictive in their eligibility requirements and more flexible with regard to the level of support provided.

FIGURE 4-1: FINANCIAL INCENTIVES FOR RENEWABLE ENERGY FROM STATES AND UTILITIES

	Grants	Rebates	Loans
Alabama	X		
Alaska			X
California		X	X
Colorado		X	X
Connecticut	X	X	
Delaware	X		
District of Columbia	X		
Florida		X	
Hawaii			X
Idaho	X		X
Illinois	X	X	
Indiana	X		
Iowa	X		X
Kansas	X		
Maine	X		
Maryland	X		X
Massachusetts	X	X	
Michigan	X	X	
Minnesota	X	X	X
Missouri	X		X
Montana	X		X
Nebraska			X
New Jersey	X	X	X
New Mexico	X	X	X
New York	X	X	X
Nevada		X	
North Carolina			X
Ohio	X		X
Oregon	X	X	X
Pennsylvania	X	X	X
Rhode Island	X	X	X
Tennessee			X
Texas		X	
Vermont	X		
Virginia	X		
Washington	X	X	X
Wisconsin	X	X	X
Wyoming	X		

Source: http://www.dsireusa.org/

FIGURE 4-2: STATES WITH PUBLIC BENEFIT FUNDS FOR RENEWABLE ENERGY

State	$ (millions)
Arizona	234
California	2048
Connecticut	338
Delaware	11
Illinois	127
Massachusetts	383
Minnesota	111
Montana	10
New Jersey	279
New York	85
Ohio	20
Oregon	95
Pennsylvania	80
Rhode Island	10
Wisconsin	22

Source: http://www.dsireusa.org/

State Environmental Funds

Thirty states and almost 500 counties are in noncompliance with the Clean Air Act due to their air quality for nitrogen oxides (NOx), sulfur oxides (SOx), and particulate matter. The Clean Air Act requires states to prepare, regularly update, and submit state implementation plans (SIPs) on how they will come into compliance with national emissions standards by certain deadlines. As of August 2004, states and counties can invest in renewable energy and energy efficiency to help them come into compliance.[6] EPA gives states a limited number of SIP "credits" for renewable energy and energy-efficiency projects that reduce nitrogen oxide emissions, the precursor to ground-level ozone. State environmental agencies often have clean air funds, or SIP funds, that invest in pollution control measures. These funds could invest in campus renewable energy projects. (See a discussion of NOx credits later in this chapter.)

A renewable energy project that can be shown to reduce NOx in the nonattainment areas of these states would be eligible for SIP credit and potentially for investment by a state or local pollution reduction fund. SIP credits and state funding can apply both to on-campus renewable energy projects and to REC purchases from local renewable energy projects that result in reduced NOx emissions within the college or university's airshed. Airsheds are large, and thus the institution could conceivably purchase RECs from a project in an adjacent state as long as that project reduced NOx emissions in its airshed.

For education and public relations purposes, the environmental benefits are often described as "the equivalent of taking xx cars off the road," or similar metrics that the general public can easily grasp. If a college or university wants its renewable energy project or REC purchase to qualify for SIP credit and state pollution control funding, it must calculate its emissions reductions far more accurately than it might otherwise have done.

State Homeland Security Funds

Most Department of Homeland Security (DHS) funds are distributed to the states. A second set of funds is distributed to the nation's 30 largest cities. Each state has a homeland security office, often located within the governor's office, that oversees the expenditure of homeland security funds. The funds are then managed in most cases by the state's emergency management agency. A portion of homeland security funds is intended to support "critical infrastructure," which includes distributed generation facilities. Renewable energy projects are a form of distributed generation, which essentially means on-site generation as opposed to large centralized power plants.

To date, there have been no state homeland security expenditures for distributed generation to the authors' knowledge. But there have been discussion and expressions of interest from state homeland security officials. In order to qualify for these funds, a renewable energy system should be designed to provide backup power generation for critical campus functions, such as medical clinics, emergency lighting, water pumping, and security. For solar or wind power projects, a battery or other storage system will be necessary so that the emergency power will be available when the sun is not shining or the wind is not blowing. For biomass power systems that rely on fuel deliveries from off campus, a backup inventory of fuel should be maintained.

EPA SUPPLEMENTAL ENVIRONMENTAL PROJECT (SEP) FUNDS

As part of enforcement settlements for alleged violators of environmental laws, EPA offers defendants the option of paying for environmentally beneficial community projects. This Supplemental Environmental Project (SEP) offsets part of the monetary penalty. Civil penalties for alleged violators often run into millions of dollars.

In 2004, EPA released *A Toolkit for States: Using Supplemental Environmental Projects (SEPs) to Promote Energy Efficiency (EE) and Renewable Energy (RE)*. It includes examples of settlements that involve renewable energy. For example, a steel manufacturer agreed to implement SEPs, including wind turbine power generation, in settlement for various violations. The SEP value was $2 million. A utility in Maryland agreed to install PV systems on three public buildings, including two schools, with a SEP value of $75,000.[7] Through a large SEP settlement for an alleged violation of the Clean Air Act, Ohio Edison Company will fund $14.4 million in wind power projects in Pennsylvania, New Jersey, or western New York. Allegheny County, Pennsylvania, will receive $400,000 to install a solar power project at a municipal building as part of this settlement.

One of the acceptable categories for SEPs is pollution prevention. Renewable energy and energy-efficiency investments fit into this category. If a college or university is in violation, it can propose an innovative SEP for an investment in renewables to EPA or its state agency. If the defendant is a corporation in the community, the college or university group can meet the corporation, state enforcement agency, and other stakeholders to propose a renewable energy project. It would also be beneficial to work in conjunction with the state energy office. The next step is to identify a portfolio of options. Based on available funds, the institution can propose a list of renewable projects and then work through the process with the involved parties to use the funds to implement projects.

▶ SAVINGS FROM ENERGY EFFICIENCY

The U.S. Department of Energy (DOE) Rebuild America program estimates that colleges and universities spend over $6 billion each year on energy and that effective energy management could save 25 percent—more than $1.5 billion in cost savings for higher education institutions. By undertaking a comprehensive energy management program, an institution could dedicate up to 25 percent of the value of its utility bill to renewable energy. It can do this without a net increase in budgetary outlays beyond what it is paying for its energy use before installing energy management projects.[8]

Many colleges and universities dedicate a portion of their energy-efficiency savings to buy renewable energy. To do this, the administration must first agree that the savings will not simply revert to the general fund. It also must decide not to apply the savings to repaying the capital cost of the energy-efficiency improvements. Instead, it must decide that the savings will flow into an account or fund dedicated to renewable energy and, if desired, further energy-efficiency improvements.

Where energy-efficiency savings are used to pay for renewable energy, they usually go for a REC purchase, not for an on-site project. On-site projects require a substantial up-front capital investment, and even large energy-efficiency savings tend to pay out over time, not all at once.

There are nevertheless some ways to use energy savings to help pay for on-site projects. Because the energy-efficiency savings flow over time, they are suited to pay the debt service on a project loan. That is, the college or university could provide a loan to the renewable energy project from its operating fund or capital fund, and the debt service could be covered all or in part by savings from previous energy-efficiency investments. The renewable energy project itself could supplement these payments with the electricity bill reductions it generates. As an alternative, the cash flow from the energy savings could contribute to lease payments for renewable energy equipment; to performance contract payments for renewable energy equipment installed on a performance contract basis; or to paying for the solar electricity generated by an on-campus installation provided under a third-party service contract (see further discussion of these financing techniques in this chapter).

If a college or university is interested in using its energy-efficiency savings to buy renewable energy, it will need to determine the dollar value of those savings. That value can either be estimated based on preinstallation engineering estimates or calculated based on actual savings measurements. Standardized procedures for undertaking these measurements and calculations are available from the International Performance Measurement and Verification Protocol (www.ipmvp.org/).

An alternative to waiting for the energy-efficiency savings to accrue before investing in renewable energy is to invest in energy efficiency and renewable energy at the same time. The predicted savings from the energy-efficiency measures can still be applied to the renewable energy purchases and investments.

A wealth of resources and information are available at the federal, state, and local levels to assist in improving energy management in facilities. State energy offices, utilities, nonprofit organizations and associations can also provide information. Good places to start are DOE/EPA Energy Star and DOE Rebuild America. Each program has resources geared specifically for colleges and universities.

Free, downloadable tools from Energy Star (www.energystar.gov) include:

- Portfolio Manager, an online tool that compares a building's energy performance to that of similar buildings nationwide. By benchmarking a group of buildings, administrators and energy managers can prioritize those with the highest potential for energy and costs savings.

- Cash Flow Opportunity (CFO) Calculator, which quantifies the cost of delaying the installation of energy-efficiency projects. The CFO's spreadsheet illustrates that installing a project results in net positive cash flow with no capital outlay through the use of financing.

- Computer power management software that can save up to $100 per year per monitor and computer by installing centralizing monitor and power management on all campus computers.

Rebuild America (www.rebuildamerica.gov) has sector-specific tools and information for colleges and universities to minimize their energy costs through effective energy management. The Solutions Center contains information, examples, resources, and links to help with planning for a variety of upgrade projects. In addition, the Department of Energy (www.doe.gov) has a range of programs and informational resources on renewable energy technology and financing.

▶ UTILITY BILL MANAGEMENT

An analysis of a university's bills may uncover overcharges and mistakes. Estimates are generally that about 1 percent of utility bills have mistakes. Companies specialize in analyzing utility bills and rate structures to ensure that the customer has the most beneficial rate class for the institution and that there are no overcharges. If the billing analysis company uncovers mistakes, these companies usually take a percentage of the utility's repayment to the end-user. If no mistakes are found, the customer doesn't pay any fee to the billing analysis company. Mistakes may include meters that are not properly working, wrong rate class, and administrative errors. Some companies will arrange for the refunds from the utility to be paid directly to the customer.

CASE STUDIES

Using Energy-Efficiency Savings to Pay for Renewable Energy

1. University of Pennsylvania

The University of Pennsylvania committed to using 5 percent wind in 2001 and entered into a 10-year contract for 10 percent wind in April 2003. At the time, this contract was the largest retail purchase of wind power in the United States, representing the electric output of 10 1.5-MW turbines, or the equivalent of 40 million kWh per year. The wind farm that supplies the university is located in Somerset, Pennsylvania.

To fund the purchases, Penn reduced its peak demand by 18 percent and 10,000 kWh. This reduction freed up funds for the costs of wind power—about $300,000 per year. The campus encompasses 138 buildings and 13 million square feet.

Student activists showed their support for sustainability activities by creating the Penn Environmental Group. The group researched, developed, and evaluated clean-energy options, resulting in the university's first wind power purchase. At the same time, there were significant energy cost increases. Students and administration looked for ways to address the costs of wind power and of "doing the right thing" for the environment. According to Barry Hilts, associate vice president for facilities, when the university decided to "marry" these two issues, it was a quantum leap in thinking.[9]

The entire wind power purchase was covered by energy management savings. To encourage energy conservation, the administration was aggressive in conservation and used a number of measures. They put the word out about specific green benefits of conservation. As a "hook," they made a commitment to purchase 5 percent wind power if energy conservation goals were met. The university held a well-publicized energy-saving drill, encouraging faculty and staff to switch off the overhead fluorescent lights between 3:00 and 3:30 p.m. if task or day lighting was sufficient. During the drill, they saved 1 MW of power (the peak demand is 62 MW). They publicized the dramatic savings. Later, computer monitors were powered down with software from Energy Star. Students did light bulb exchanges in residences halls to replace inefficient incandescents.

The administration aggressively manages peak load because the university is on a ratchet charge. On high-humidity, high-temperature days, phased conservation is instituted for the whole campus. A peaking diesel generator can shave 2 MW of demand. They gradually raise the temperature in their chilled water loop. These efforts eliminate the need for turning on additional chillers, saving 1 MW per chiller. The EMS system monitors demand. A thermal storage unit makes ice during the night, which is then used to supplement the chillers during daytime peak. Air handlers shut down on a 30-minute roving schedule (except in research and other selected buildings). One in four elevators can be shut down, and lights on garages can be turned off. The facilities staff works with building administrators to monitor energy use, ensuring that lights and dedicated air-handler units are turned off in empty conference rooms.

Penn achieved its goals of encouraging conservation and becoming a national leader in the green energy market. Some of the additional benefits have been support from the research faculty, peer-to-peer exchange, student involvement, and positive public relations.

2. SUNY University at Buffalo

The State University of New York University at Buffalo's aggressive energy conservation program has resulted in savings of more than $9 million per year. The university used part of these energy management savings to make a purchase of clean, regionally generated wind power.

According to energy manager Walter Simpson, UB has reduced the total energy consumption on campus by 30 percent or more through energy conservation. The university, which has 100 academic buildings and a $20 million annual energy bill, has worked extensively with energy service companies (ESCOs) to institute energy management projects through performance contracts. These projects have paid for themselves and created a new cash flow with no up-front funds. Between 1993 and 1997, UB worked with an ESCO and completed $17 million in comprehensive energy management projects. Simpson estimates that these projects would have taken between 10 and 15 years without the partnership with the ESCO. This initiative produced more than $3 million in annual savings.[10]

UB used a ramping-up approach to green power purchases. President William Greiner approved a sustainable energy policy in May 2000. In 2002, the university became the first in the SUNY system to buy a portion of its electrical power from a commercial supplier of wind-generated electricity when it purchased the output of 1.5 MW from the Fenner Wind Farm in Fenner, New York. UB increased its purchase in 2003 to 8 million kWh of wind energy from Wethersfield Wind Farm in Wyoming County, New York. In 2004, UB increased its purchase again to 12 million kWh. The 2003 and 2004 purchases make UB the largest purchaser of wind energy of any New York state agency. Their purchase represents 4 percent of the university's electric consumption for 2003 and 6 percent for 2004 and 2005. The university is now considering proposals to install wind turbines on campus and to use regional biomass for heating needs.[11] UB has been very supportive of Governor George Pataki's Executive Order 111, which directs SUNY and other state facilities to increase their use of renewable energy. UB is on track for achieving the mandated requirements.

▶ GIFTS FOR ENVIRONMENTAL EFFORTS

Colleges and universities have received gifts and endowments in recognition of their commitment to environmental and sustainability efforts. Although these gifts are not directly tied to a renewable energy commitment, an institution is more likely to have increased credibility if it is powered by clean energy.

These are some examples:

- In 2004, Arizona State University received a $15 million gift from a philanthropist to create an International Institute for Sustainability. This seed money is establishing an umbrella organization for the university's sustainability activities.

- The College of the Atlantic received a $1 million gift to develop and expand its sustainability initiatives at the same time as it committed to 100 percent wind power.

- Ithaca College received a $7 million gift in 2004 from Dorothy Park, president of the Park Foundation. The gift—the fifth largest in the college's history—will fund additional green building elements in the School of Business's new facility, scheduled to be completed in 2008. The $14 million building will be designed to achieve LEED Platinum certification (the highest level) from the U.S. Green Building Council. As part of the design, passive solar units are planned on the back of the building.[12] Ithaca College will also be studying the feasibility of installing wind turbines on South Hill, part of the campus.[13]

CASE STUDY

College of the Atlantic—Gift for Wind Power

Sustainability has been a core value of College of the Atlantic (COA) in Bar Harbor, Maine, since its inception in 1969. COA was the first college to offer a degree in human ecology—the degree that all its graduates receive. COA received a $1 million endowment from trustee Henry Sharpe and his wife Peggy. The Sharpe Fund for Organizational Stewardship also provides a $250,000 challenge matching fund, so that the total gift can reach $1.5 million. Through this endowment, the college has created an Office of Sustainability and hired a director. It has also established a lecture series related to environmental stewardship and ecological entrepreneurship and will develop new classes around these areas of interest.

The college has been addressing energy issues by making energy efficiency lighting upgrades in recent years. All new buildings will have passive solar systems. In addition, COA is part of the State of Maine's STEP-UP Program run by the Maine Department of Environmental Protection. COA committed in its STEP-UP agreement to reduce total energy use by 30 percent, GHG emissions by 30 percent, and to contract for 100 percent of its electrical energy use from a renewable source, all by November 2006.

In celebration of Earth Day 2004, COA was the first college in the United States to enter into a long-term contract to purchase 100 percent of its electrical energy needs from wind power. The college's commitment has obvious environmental benefits, and by contracting with a new local wind power corporation, COA is helping to bring wind energy development to Maine with the Redington Wind Farm Project. Additional benefits include showing students that COA is serious about its values and ultimately strengthening recruitment efforts.

The board of trustees scrutinized the fixed-price contract for wind power. No prepayment was required, and board members determined that the pricing structure provided less risk than oil pricing. The contract was signed before the dramatic increase in oil prices. Subsequently, the wind power corporation changed suppliers. The contract with the new supplier is one-half fixed price, and the other half increases at half the rate of inflation. Although the initial costs are slightly higher than anticipated, COA expects that there will be long-term savings over the 10-year contract. COA received local, state, and national coverage of its environmental initiatives in newspapers from the *Mount Desert Islander* to the *New York Times*.

In 2004, COA was the only independent college in the nation to receive a Green Power Leadership Award from the U.S. Environmental Protection Agency and the U.S. Department of Energy. COA has also received two National Wildlife Federation Campus Ecology Recognition Awards for its renewable energy and sustainability efforts and for its zero-waste graduation.

- The Wisconsin Public Service Foundation provided the University of Wisconsin at Oshkosh with a $250,000 grant to design a LEED building on campus. The university is also purchasing $25,000 in renewable power from the utility—a mixture of 50 percent wind, 40 percent landfill gas, and 10 percent farm waste. At 4 percent, the university is now the largest green power purchaser in the state.

CASE STUDY

Cornell University—Using Gifts to Capitalize a Renewable Energy Fund

The Cornell Solar Fund, established by a Cornell student and managed by the university's facilities office, is capitalized through gifts from students, professors, alumni, and the friends and family of the founding student. The fund provides cost-share subsidies in the form of grants to cover a portion of the cost of on-campus solar energy projects. The fund had a little over $12,000 by the end of 2005 and had not yet made any disbursements, although two projects were under consideration.

During the fund's establishment, consideration was given to how it might become a financially sustainable investment vehicle. No charter or formal investment guidelines were established, partly for tax reasons, but the intent was for the fund to invest in projects that have a high degree of visibility. The investment size is based on the simple payback of the project. Cornell facilities management uses a seven-year investment horizon, so the fund provides a level of subsidy to bring a project's payback down to seven years.

For administrative and tax reasons, the Solar Fund is subsumed in the Cornell Fund, the university's prime fund-raising vehicle.

▶ GENERAL FUND

Special allocations from general funds are a common source of capital for renewable energy investments. Often, the funds are provided to projects as loans and are then repaid from energy bill savings or government rebates.

With funds from its operating budget, Catholic University of America agreed in 2002 to purchase the full output of one 1.5 MW wind turbine (12 percent of its load) for $72,000 for five years. It has been able to cover the additional costs with surpluses developed through various energy conservation projects in past fiscal years. This purchase was consistent with its mission of stewardship and a good fit with its overall plan of responsibility and respect for the environment.[14]

California State University Northridge (CSUN) dedicated a portion of its general fund to capitalize an energy reserve fund, which is used for energy efficiency and renewable energy. The energy fund is not required to reimburse the general fund, although any rebates earned from energy fund investments go to the general fund. The revenues earned from rebates on CSUN's first solar project helped finance its second solar project, but the rebate revenues went to the general fund, as will rebates from the second project.

In 2005, Harvard University created a $100,000-a-year renewable energy fund that will help to increase renewable energy initiatives and projects on campus. It is a three-year dedicated fund. Harvard also established a $3 million revolving loan fund called the Green Building Loan Fund, which provides financial incentives for high-performance building designs and technologies in renovations and new buildings.[15]

CASE STUDY

Duke University—Combining Student Fees with Energy Savings and Matching Funds

Duke University uses a combination of energy savings and other funds for its renewable energy purchases. The Duke Green Power Challenge was a student-driven initiative. In 2002, the students started by counting the incandescent light bulbs in university-owned fixtures that had not been switched to more efficient compact fluorescents. The students presented the potential savings to the administration. To account for the labor savings—since compacts last 10 to 12 years and don't need to be changed as frequently as incandescent bulbs—they adjusted the figures by a factor of 10.[16] In 2003, the administration agreed to use a portion of the cost savings to match student purchases of wind power up to $25,000.

The university's renewable energy provider trained 40 students on wind power and how to sign people up. In 2003, the provider sold $25 blocks—the estimated power usage per dorm room per semester. A block was estimated to be 1,850 kWh. In three weeks, 500 blocks were sold; some students purchased two blocks, or two semesters' worth. This significant student effort led to subsequent purchases. Duke is now purchasing green power for approximately 10 percent of its load.

In Duke's Pratt School of Engineering, the actions were taken for educational purposes, and the funding came from an academic budget. The Fitzpatrick Center was two points away from a LEED Silver rating. This 320,000-square-foot lab opened up in November 2004. Students were involved with the construction plans. Since this was post-construction, there were few options for getting to silver. The students suggested the option of renewable purchase and the administration agreed to pay. This is to be an ongoing purchase and the building is to be used as an educational tool. The whole building is considered a "classroom." Data sets from building operations are being used by professors in three classes.

Also, the School of Environment is purchasing 8 million kWh/year for two years. Duke now has a master contract with its wind power supplier. As other parts of the university want to buy wind, the purchasing mechanism is in place that simplifies the procurement process.[17]

▶ STUDENT FEES

Students throughout the country have organized referendums to enable student fees to be used for renewable energy purchases. Some referendums have proposed using a portion of current student fees, while others have proposed increasing fees. This approach has worked successfully at many colleges and universities. Generally, students have voted in strong favor of increasing student fees in order to purchase clean energy.

The first university to hold a referendum was the University of Colorado at Boulder. In 2000, there was a student-led initiative to increase student fees by $1 per semester for four years. The referendum passed by a margin of 5 to 1, with the highest turnout in the university's history. In 2003, 92 percent of student voters approved a $2 per semester increase so that three student buildings could purchase 100 percent renewable energy. This method has now become widespread (see figure 4-1).

FIGURE 4-3: RENEWABLE ENERGY STUDENT-FEE REFERENDUMS

Appalachian State University: Eighty-one percent of students voted to approve a $5 per semester increase in student fees.

Auraria Campus (Metropolitan State College of Denver, Community College of Denver, University of Colorado at Denver): On Earth Day 2004, 95 percent of students approved a $1 per semester fee to purchase wind power. Of these funds, $100,000 will be used for a solar demonstration and the remainder will be used for a wind purchase.

Connecticut College: Ninety percent of student voters agreed to pay an extra $25 a year in tuition for a renewable energy purchase of 17 percent in 2001. In 2002, this amount increased to 22 percent. In the 2003–2004 academic year, the college procured 44 percent green power, including a blend of wind, biomass, and low-impact hydro-power. Seventy-six percent of students voted.

Eastern University: In 2003, through a student initiative supported by the administration, the university agreed to purchase 37 percent locally generated wind energy. The cost was split in half between the administration and the students' fees.

Evergreen State College: In a campus referendum in 2005, students voted 10 to 1 to pay an extra $1 per class credit per quarter to procure clean power.

Harvard University, Kennedy School of Government: Fifty-eight percent of students voting approved a $5 wind energy surcharge per semester's tuition bill to procure 100 percent clean energy. Later, the administration announced that it would pay for the purchase through an administrative budget.

Humboldt State University: Eighty-five percent of students voted to add $1 to their tuition for green power.

Lewis and Clark College: Students created a voluntary $25 per quarter check-off in the student registration materials. In 2004, contributions resulted in a 400 MW-hour purchase of renewable energy.

Smith College: Seventy-five percent of students voted on a clean energy referendum in favor of pursuing renewable energy.

Tufts University: Eighty-eight percent of student voters were in favor of Tufts Environmental Consciousness Outreach's referendum to institute wind power at an average cost of $20 per student per year.

University of Denver: Seventy-eight percent of students voted in favor of a student fee for renewable energy.

University of North Carolina, Chapel Hill: Seventy-four percent of student voters approved a $4 per student per semester fee for clean energy.

University of Tennessee, Knoxville: In 2004, student voters agreed to an $8 per semester increase to purchase alternative energy.

University of the South: The Student Assembly approved a resolution committed to phasing in green energy. The university has agreed to purchase 5 percent green power in 2005, 10 percent in 2006, and 15 percent in 2007 and thereafter.

Western Washington University: Eighty-five percent of student voters supported an initiative to purchase renewable energy for a quarterly fee of no more than $19 per student. Starting in 2005, the university began purchasing 32 million kWh of clean energy, which represents 100 percent of the load.

Source: EPA Green Power Partnership and www.envirocitizen.org

Student fees generally range from a few dollars to approximately $25 per semester based on students' interests. After a green power purchase funded through student fees is in place, the administration often recognizes the strong commitment and augments funding for additional clean power. These efforts are sometimes developed and implemented by a collaboration of students and faculty.

Some institutions use a "university challenge" to obtain funds for renewable energy purchases. Typically, the administration recognizes the strong desire of students and faculty to improve the institution's environmental footprint. To improve energy efficiency and free up funds for a clean energy purchase, students and administration organize an energy reduction "challenge."

Students often take a lead in organizing the challenge and spread the word to encourage all members of the campus community to be more energy conscious and minimize their energy consumption. They educate other students and staff, develop brochures and information resources, and knock on doors. As a result of this strong demonstration of student support, some administrators take the next step and provide funding for the clean power purchase.

▶ DEBT

A college or university's ability to use debt to finance renewable energy projects will depend on its debt capacity and the limitations of its charter or statute. Many colleges and universities have some level of unused debt capacity. Energy and infrastructure investments are usually well within the norm as eligible uses for debt. Policies governing debt liability may differentiate between revenue-generating facilities and nonrevenue-generating facilities. Revenue-generating projects such as renewable energy projects are often considered to be good candidates for debt financing.[18]

Often, debt financing must be secured by a security interest in the renewable energy project or supported by other collateral, such as an acceptable surety or equivalent guarantee in the amount of the lender's financing. Such conditions will come into play particularly when the college or university establishes a new entity to build, own, and operate the renewable energy project, thereby shielding the institution from recourse by the lender. One reason for state colleges and community college districts to establish private entities to own and operate energy projects is to avoid the debt limits placed on them by statute. For example, Arizona public institutions cannot enter into agreements that involve more than two years of repayments. The president of Arizona State University has authority to extend the limit, which he did in the case of a clean energy bond the university issued. New Jersey's public colleges are limited to five-year loan terms.[19] They have therefore been unable to participate in the state's low-interest loan program for clean energy projects.[20]

State and community college districts have used bond issuances to include renewable energy components. In 2004, the Kern County Community College District in California used $9.2 million out of a $180 million, 25-year general obligation bond to cover 68 percent of the cost of a 1-MW solar PV system installed at Cerro Coso Community College's Ridgecrest campus. The solar project headlined the bond issuance and was popular with voters, according to college officials.[21] Arizona State University financed the entire cost of its 30-kW solar PV project as a component of a bond issued for a package of energy-related projects on the ASU campuses.

▶ PRIVATE EQUITY INVESTORS

Private investment is applicable mainly to larger wind power projects, which can be profitable for the college or university. To attract investors, the institution may establish a project entity, such as a limited liability company (LLC) or a corporation, which provides protection from liability and access to tax incentives. The college or university, as project sponsor, negotiates a term loan for the purchase of the wind turbine, associated equipment, and possibly for construction. It will have to negotiate the interest rate, term, loan amount (based on the bank's desired percentage of equity), and amount of collateral required, if any. The institution could decide to be the sole equity holder, but the economics may be too costly. It may be best to solicit outside investors who will not usually be interested in a project so small without the availability of renewable energy tax credits.

The type of investors to solicit are those with profitable businesses and a sizable tax exposure who thus have a tax "appetite" that they would like to offset. The return on their equity investments comes in part from the tax benefits—both the federal production tax credit and depreciation on the wind turbine. They also earn a return from the electricity and REC sales and from the residual value of the turbine sold at fair market value to the college or university at the end of a given contract period. When the contract period ends, ownership of the wind turbine is "flipped" from the private investors to the institution. The institution pays the residual value of, say, $150,000 on a 10-year-old $2 million wind turbine and then derives 100 percent of the benefits—power sales, REC sales, and salvage value—from that point forward.

The short-term monetary benefits to the college or university will be small. The overall benefit of this approach is that it has attracted an equity investment in a renewable energy project that would not otherwise have happened and that the college or university will eventually own outright. There may also be ways for the college or university to negotiate a small percentage of the power sale revenues or tax credit revenues for itself during the contract period.

▶ LEASES

A college or university can lease renewable energy equipment under operating leases or capital leases. There are two main differences between the two lease types. A capital lease assumes the institution will own the equipment at the end of the lease term, and thus the lease payments will be somewhat higher than under an operating lease. Since the institution will be the owner, a capital lease assumes it is the owner for tax and depreciation purposes. Since colleges and universities do not pay taxes, this benefit cannot be used. Under an operating lease, the leasing company (lessor) will get those benefits, thereby making the college or university's payments under an operating lease lower still.

Leasing is most applicable to certain kinds of biomass power generation equipment and solar PV modules, but also may be applicable to small wind turbines. Unlike loans, leases will cover only equipment that, if necessary, can be repossessed and released or sold on a secondary market. Leases generally will not cover batteries, controls, and other ancillary equipment, especially if the leasing companies believe the equipment's residual value will be too low compared to its original price. Most leasing companies specialize in a limited range of equipment and will not provide leases for equipment with which they are unfamiliar. Most are unfamiliar with renewable energy technologies. However, some specialized leasing companies lease PV modules and other specialized equipment if they are familiar with its resale value and the characteristics of its secondary market. Many of the generators and boilers used for biomass combustion, including gas turbines and fuel cells, are standardized products used for conventional fuels as well. Thus, there is a relatively large secondary market for them, and leasing companies will be more comfortable providing leases for them.

Most leasing companies will be flexible with payment schedules and will work to accommodate their clients' cash flow. The ideal way to structure a lease is to match the college or university's lease payments to the projected utility bill savings so that there are no out-of-pocket financing costs. However, this approach will likely require a long-term lease in order to reduce the size of the lease payments. Most lease contracts run only three to seven years, but they can be up to 30 years long, which may be an appropriate choice if cash flow is a paramount concern.

Leasing contracts typically include these terms:

- The leasing company (the "lessor") will remain the equipment owner. The client (the "lessee") will acquire the right of temporary possession and use.

- The lessee must pay one or more lease payments when the lease is signed and the lessee obtains possession of the equipment; subsequent payments are usually made at periodic intervals.

- The lessor may or may not recognize a salvage value in calculating leasing payments.

- Often the lease cannot be canceled, and if it is canceled, a substantial penalty may be imposed.

- The lessee typically is responsible for property taxes, insurance, and repairs not covered by the warranty, although this term is negotiable.

- When the lease period ends, the lessee has the option to purchase the equipment, renew the lease, or return the equipment to the lessor.

▶ PERFORMANCE CONTRACTS

Performance contracts can be used as a financial instrument to spread the cost of the renewable energy installation over a multiyear contract period, much like a loan or a lease. The difference is that the performance contract's monthly payments are made to the equipment vendor/installer and can be linked to the estimated utility bill savings resulting from the installation's performance. Ideally, the college or university's monthly payments are equal to or less than the monthly utility bill savings, thereby providing positive, or at least neutral cash flow during the contract period. After the contract period, the college or university gets the full utility bill savings. Performance contracts shift the risk during the project's startup to the private sector provider. This risk shifting is one of the major benefits of entering into longer-term performance contracts when a project's long-term performance is uncertain.

Performance contracts are common for energy-efficiency installations when the payback is relatively short. It is less common for renewable energy projects. Most renewable energy investments have longer paybacks than most energy-efficiency investments. Thus, if a renewable energy project's monthly payments are to be paid out of utility bill savings, the performance contract's term will have to be inordinately long, perhaps 20 years or longer. Most performance contract providers, including renewable energy vendors, are unlikely to agree to such a long contract term.

The college or university can always agree to a shorter contract period and simply pay the extra amount per month. As an alternative, the institution can combine the renewable energy installation with another energy project of shorter payback, such as an energy-efficiency project or a combined heat and power (CHP) project. Pierce College combined its 191-kW solar photovoltaic installation with a CHP project. This combination increased the total project cost but reduced the net payback of the two projects, allowing for a shorter contract period on the performance contract.[22]

State colleges and universities may face regulatory restrictions on their ability to engage in performance contracting. For example, Arizona State University was interested in a performance contract for its solar PV system, but faced the procurement code issued by the Arizona Board of Regents, which limits contract terms to a maximum of five years. The repayment period for the PV system was approximately 11 years. Fortunately, the procurement code allows the overriding of the five-year limit if the president of the university approves a longer term in writing. The president of ASU agreed to the longer contract term in this case.[23]

In some cases, it may be advantageous to simply sign a one- or two-year performance contract. A shorter-term contract is less costly, but it serves as a guarantee that the project will perform correctly at least during its initial phase of operation when startup problems may occur. This approach removes the financing aspect of the performance contract because now the college will have to come up with some other form of financing.

▶ THIRD-PARTY SERVICE MODEL

The third-party service model, used mainly for solar PV installations, allows colleges and universities to use solar electricity generated from solar installations at their facilities for no upfront cost by entering into a "solar services" contract. A third-party company designs the system, provides the equipment, and arranges for the installation. It owns, operates, and maintains the system and accepts the performance risk. The college or university pays only for the electricity generated from the system at its facility. It does not have to come up with an upfront down payment. It only enters into a services agreement to buy electricity generated, in the same way as it would sign a typical power purchase agreement. The third party also arranges for liability and force majeure insurance. The college or university faces no performance risk. It simply provides access to a rooftop or other suitable location for the solar system and agrees to purchase electricity from the system at a long-term fixed price—usually equal to or less than the current price for conventional electricity—for a 10- to 20-year contract period. The capital for this investment comes from investors seeking to reduce their tax exposure by qualifying for available state and federal solar tax credits and other available incentives. The service provider also may take title to any RECs or other emissions credits that accrue to the project.

The main service provider of this type is SunEdison, LLC of Baltimore. The company has an arrangement with BP Solar, which provides and installs the solar modules. SunEdison operates in states with significant solar subsidy programs, including California, Connecticut, and New Jersey. It will soon be entering into markets in Nevada, Arizona, Colorado, and Pennsylvania. The first higher education institutions to pursue the third-party service model are Stevens Institute of Technology in New Jersey and the California State University campuses of Fullerton, Dominguez Hills, and San Luis Obispo. The California campuses signed 20-year agreements with SunEdison.[24] Stevens Institute signed a 25-year agreement. It will buy the output of a 125-kW solar installation at 9 cents per kWh. The college's average electric rate is currently 8.3 cents per kWh, but that rate is expected to rise higher than 9 cents some time within the next 25 years, at which point the solar installation will begin saving money.

Sterling Planet, a large marketer of RECs, used the third party service approach to install two small solar PV systems at Florida State University. Sterling Planet sells the electricity to the university and sells the RECs to the City of Tallahassee.[25]

At the end of the service contract, a college or university has three choices for how to proceed. It can sign another long-term agreement with the third party to continue to receive the service; end the agreement and have the solar installation removed; or purchase the system at fair market value.

The third-party service model is starting to be applied to anaerobic digestion-biomass power generation projects. The third-party company signs contracts with farmers or agricultural colleges to design, build, own, and operate projects on site at dairies with no financial outlay by the farmer or the college (i.e., the dairy owner). The power is sold either to the dairy owner or the local utility, and compost is sold to nurseries. In some cases, the compost can be the higher revenue producer and will boost the returns on what is essentially a biomass power generation. To date, the biomass third-party model has not been applied at any agricultural college or university.

▶ ELECTRICITY SALES

Selling the output of a campus renewable energy project can be the main vehicle for covering the cost of the project. This has been the case, for example, for the wind turbines installed by Carleton College and Iowa Lakes Community College (see chapter 3). Most campuses consume all the renewable energy they generate on site, but if the system is large enough or if campus demand drops during the summer or other times of the year, it is possible to sell the excess power to the local distribution company or to other electricity consumers who are interested in buying green power.

Regulations allowing the sale of excess power to a utility are determined on a state-by-state basis by public utility commissions. Most utilities oppose it. In general, the more reliable the renewable energy generation is, the higher the price the utility will pay for it. Thus, intermittent sources like solar and wind will generally bring a lower price than biomass, hydro, or geothermal, which can operate 24 hours a day. However, solar may be able to attract a good price since it generates electricity during daylight hours when most utilities face the greatest demand for their electricity and thus need to have generation available just for those peak hours.

Many electricity consumers interested in buying green power may prefer to buy it from the college or university rather than from other green power suppliers. The college or university can either sell the electricity to these customers directly or it can sell the customers only the RECs while selling the electricity to the utility. This latter approach may be preferable because the utility will likely charge the college or university a fee to send or "wheel" the electricity to the customers. State utility commissions regulate wheeling fees, but the fees may add too much to the price of the green power. The utility may also be unfriendly toward the notion of the institution taking away its customers. Larger colleges and universities that purchase a large amount of electricity from their utility will have greater leverage to negotiate innovative schemes like this.

It is possible simply to provide the excess electricity to the grid free of charge. For example, when the solar PV panels at Oberlin College in Ohio produce more energy than is needed by the building on which they are installed, the excess power is donated to the local utility, displacing some coal-fired power production. Although Oberlin does not calculate the resulting reduced air emissions from the power plant, emissions reductions can be calculated, commoditized, certified, and sold as either renewable energy certificates, eligible NOx reductions, or both.

▶ SALES OF RECS AND OTHER ENVIRONMENTAL PRODUCTS

Along with the generation of electricity, campus renewable energy projects generate environmental benefits, which can be commoditized and sold to interested buyers such as electric utilities, brokers, or other institutions or individuals who want to offset their consumption of "brown" power with some "green" power. The environmental commodities are renewable energy certificates (RECs), measured in megawatt-hours, and NOx credits, measured in tons.

Renewable Energy Certificates (RECs)

College and universities in most states can sell their RECs in the voluntary market to anyone in the nation who wants to buy them. They can also sell them to brokers or REC aggregators, which are private firms that purchase RECs from renewable energy project owners and sell them for a profit to interested buyers (see appendix B for a list of brokers). A college or university interested in building a renewable energy project should contact these brokers to see what price they will pay for RECs and what conditions apply to the sale.

Colleges and universities in states with RPS requirements may be able to sell their RECs on the compliance market to in-state utilities seeking to comply with the RPS. Each RPS defines the eligible renewable energy sources and the geographic area in which the RECs have to be created. However, it should be noted that utilities often have a choice from whom to buy their renewable energy, either directly or in the form of RECs, and just because a college or university is located in an RPS state does not mean utilities will necessarily buy their RECs. In fact, utilities typically seek to comply in the easiest and least costly manner, which means buying a large amount of renewable energy from large providers, not a small amount from many small providers like colleges and universities.

Colleges and universities can sell their RECs either through payment-on-delivery contracts or forward contracts. Under a payment-on-delivery contract, the REC buyers agree to buy a set number of RECs at a set price over a set time period. The buyer makes annual, quarterly, or even monthly payments to the REC supplier as the RECs are generated.

Under a forward contract, the buyer agrees to pay upfront for a set number of RECs that will be generated in the future. The RECs bought that are not yet generated would have a future "vintage." Another term for this arrangement is a prepaid purchase contract.[26] The obvious advantage of this approach is that the college or university can receive a prepayment to pay for the construction of its renewable energy project. California State University, Hayward, sold five years' worth of RECs from a solar project on a prepaid purchase contract basis. Revenue from the $600,000 sale is being used to help cover the debt service on the loan that financed the project (a 1-MW solar PV installation in 2004).[27]

Although forward contracts may be preferable, payment-on-delivery contracts still have significant value because the anticipated revenue can be dedicated to servicing the project's debt. When the college or university has chosen to establish a new organizational entity, such as an LLC, to build, own, and operate the renewable energy project, lenders will have more comfort in being repaid if they can see a steady income stream from REC sales that can be dedicated to servicing the debt. In fact, they may ask to receive that income stream directly for debt service purposes, without it first going to the college or university. As an alternative, a lender and the college or university could agree to the establishment of an escrow account for the REC receipts. The account would be managed by a third-party custodian and would be used for the sole purpose of servicing project debt until the debt is retired.

The main challenge with selling RECs, whether as payment-on-delivery contracts or forward contracts, is that it is often difficult and time-consuming to identify and close sales contracts with potential REC buyers. The most obvious buyer is the local utility. If the utility is not interested, the college or university has its work cut out. It will need to find buyers for all the RECs that the renewable energy project generates each year, so selling RECs will be an ongoing activity. If the college or university cannot identify one or two willing buyers, it may be cost-prohibitive to try to sell all the RECs, plus establish a system for tracking the REC sales, invoicing the buyers, and marketing newly generated RECs.

An alternative to searching for REC buyers is to sell the RECs to a brokerage firm or private investment fund. These firms act as clearinghouses between sellers and buyers of RECs and other environmental products. Selling to these firms is convenient, but they make their money off the differential between the buying and selling price, so they will tend to offer as low a price for a college's or university's RECs as possible. Also, they may not be willing to buy REC forward contracts and may only be willing to buy the RECs through payment-on-delivery contracts.

An idea for the future is for colleges and universities to work together to form their own REC clearinghouses. Many that are not installing their own renewable energy systems are interested in purchasing RECs, typically from either their electricity supplier or REC broker. These institutions may prefer to buy their RECs from other colleges and universities. A network or clearinghouse among the REC-buying and REC-selling institutions would benefit both groups. Without the brokerage intermediary, the selling institutions would earn a higher price for their RECs, while the buying institutions would pay a lower price for their RECs. A small administrative fee would be needed to cover the costs of managing such a clearinghouse.

NOx Emission Allowances

Under certain circumstances, the RECs accruing to a college or university from its renewable energy project can theoretically be sold to state and local governments as part of those governments' efforts to comply with the Clean Air Act. The act requires states to submit and regularly update state implementation plans (SIPs), which lay out the state's and local government's actions for controlling pollution. Local governments contribute to the establishment of the plans and also carry out part of the implementation. Under an August 2004 EPA guidance, state and local governments can receive emission reduction credit in their SIPs for purchases of renewable energy that reduce nitrogen oxide (NOx) emissions and help achieve attainment of the National Ambient Air Quality Standard for ozone.[28] In order to be sold in the NOx reduction market, the REC must be converted from an energy commodity measured in kilowatt-hours to a pollution commodity measured in tons. Once this transaction has occurred, the original commodity—the REC—can no longer be used or traded under current market practices, as that would be considered double counting.[29] Once the state or local government buys the RECs, it "retires" the equivalent number of NOx allowances.

The NOx reductions that result from renewable energy generation are eligible for SIP credit in only certain states and regions (figure 4-4). Three conditions must be met. First, the NOx reductions must occur in those states or regions of the country that are part of the Clean Air Interstate Rule (CAIR) where NOx emissions are regulated under a cap-and-trade system, which is explained below. The states where NOx emission allowance trading may take place are located in the eastern part of the country. Second, the states or regions must be in nonattainment status for ozone. The ozone nonattainment areas are generally located in and around large metropolitan regions in the East and Midwest. Third, the state must implement a NOx set-aside for renewable energy such that actual NOx emission allowances are pulled from the state allocation normally awarded to utilities. This, in effect, pulls down the emission cap. Under this scenario, the state may require that the NOx emission allowances be retired (rather than sold) to ensure NOx emissions have been removed from the market. Thus, it is mainly colleges located in midwestern or eastern nonattainment zones that can consider selling their state allowances to state and/or municipal governments.

As of early 2006, no college or university has yet sold NOx reductions to a state or local government. To the authors' knowledge, none have tried. Colleges and universities in these eligible locations can explore this possibility with their state environmental agencies and/or regional EPA offices. Although the EPA guidance lays out recommended procedures for how renewable energy can receive SIP credit, the approval of SIP credits is made by regional EPA offices, which are allowed some latitude in how they interpret the guidance.

Colleges and universities that are located in ozone nonattainment, cap-and-trade areas face the following regulatory situation. Under the Clean Air Act, the U.S. EPA and state governments have set emission caps (in tons) for particular pollutants (e.g., NOx) in a

FIGURE 4-4: CAP-AND-TRADE AREAS UNDER THE EPA CLEAN AIR INTERSTATE RULE

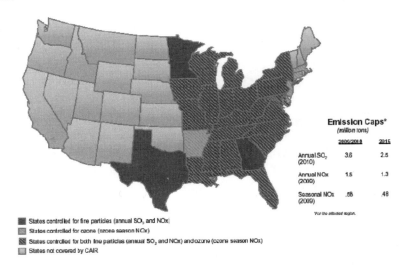

Emission Caps*
(million tons)

	2009/2010	2015
Annual SO₂ (2010)	3.6	2.5
Annual NOx (2009)	1.5	1.3
Seasonal NOx (2009)	.58	.48

For the affected region.

■ States controlled for fine particles (annual SO₂ and NOx)

 States controlled for ozone (ozone season NOx)

▨ States controlled for both fine particles (annual SO₂ and NOx) and ozone (ozone season NOx)

 States not covered by CAIR

Source: U.S. Environmental Protection Agency, Washington, DC

particular sector (e.g., electric power generation). They also distribute allowances that permit emissions of a specific amount of a pollutant over a specified time period (e.g., calendar year). The power producers meet the requirement either by reducing their own emissions or by buying allowances. Power producers that reduced their emissions below their allowances can sell their extra allowances to other power producers. State regulators can "set aside" some allowances for wind and solar energy projects. The owners of these projects, such as colleges and universities, would likely be prohibited from selling the allowances to polluters because that would not reduce pollution. More likely is that the state or local governments would buy the NOx reductions from the renewable energy projects and "retire" the allowances, thereby permanently removing the equivalent amount of NOx pollution.

Energy efficiency and renewable energy set-asides for NOx trading programs are in effect or under development in Connecticut, Illinois, Indiana, Maryland, Massachusetts, Michigan, New Hampshire, New Jersey, New York, Ohio, and Virginia. Georgia will soon be considering whether it too will offer a renewable and energy-efficiency set-aside program for NOx allowances. The landmark case in which renewable energy was purchased for SIP credit took place in 2004. Montgomery County, Maryland (a suburb of Washington, D.C.) led a consortium of municipalities and the local water district in buying 38.4 million kWh per year of wind RECs. At the time, this purchase was the largest local government wind purchase in the United States. The county issued an RFP for local wind RECs and purchased the RECs from a local utility subsidiary that in turn purchased them from a REC marketer.[30] The RECs were converted to approximately 40 tons of NOx emission reductions, and the county retired an equivalent number of NOx allowances. The EPA approved the purchase pursuant to its August 2004 guidance. Other states and local governments are ramping up to be able to make REC purchases for SIP credit a little easier (through set-asides, for example).

FIG. 4-5: EPA-DESIGNATED NONATTAINMENT AREAS FOR OZONE AND PARTICULATE MATTER, 2004

Source: U.S. Environmental Protection Agency, Washington, DC

Carbon Credits

Although there is increasing interest in carbon credits and carbon markets, there is still extremely limited scope in the United States for carbon credits to play a significant role in helping finance renewable energy projects. Because carbon is not yet regulated in this country, all carbon reduction efforts are voluntary.

Many observers believe carbon will be regulated at some point and that a carbon market will emerge and develop, possibly with some form of carbon emission trading that is now under way in Europe. One of the possibilities for the U.S. market will be the certification of certain carbon reductions that took place prior to the establishment of the carbon regulations. These "grandfathered" reductions would potentially be eligible for sale in a U.S. carbon market. Thus, even if a college or university renewable energy project cannot currently expect to sell or trade its carbon emission reductions (CERs), it should nevertheless quantify those CERs in the event that they become eligible under possible grandfathering rules. The quantification can employ one of several approaches, each of which uses different techniques to establish a baseline and calculate emission factors. The major approaches are those of the EPA and the Chicago Climate Exchange. Some states have a quantification procedure as well.

Colleges and universities that have offices abroad or international affiliations may want to investigate the possibility of earning carbon credits for renewable energy or energy-efficiency projects undertaken at those offices or affiliated facilities.

▶ MINIMIZING THE COST OF RENEWABLE ENERGY PURCHASES

Blending: Increasing the Proportion of Less Expensive Green Sources

To minimize costs, some institutions choose a blend of green sources for their product mix. Solar energy and wind are the "cleanest" in terms of emissions but also more expensive than other forms. By mixing in less expensive renewable sources, such as landfill methane, colleges and universities are able to purchase higher percentages of renewable energy. This mixture often includes older sources, such as hydropower.

A group of colleges and universities in New Jersey procured clean power in conjunction with several state agencies. The mixture consisted of half wind power (more expensive) and half landfill gas (less expensive). The mixture for this purchase included landfill gas so that the overall percentage could be increased to 15 percent.

Aggregation

Colleges and universities can aggregate for better pricing from green power suppliers. The Pennsylvania Consortium for Interdisciplinary Environmental Policy (PCIEP) is an organization of the Commonwealth of Pennsylvania's Department of Environmental Protection and Department of Conservation and Natural Resources. Through PCIEP, more than 40 Pennsylvania colleges and universities are part of the aggregate purchasing arrangement to buy wind and are purchasing at a lower rate than what their individual rates would be.

Six colleges and universities in New Jersey have aggregated their purchase of fuel cells to obtain better pricing. Some New Jersey institutions are also aggregating in conjunction with state agencies to obtain better pricing for green power.

At present, the largest purchase for a university system of renewable energy through RECs in the United States is an aggregate purchase by the University of California (UC) and California State University (CSU) systems. In June 2005, a contract was signed for a 15 percent purchase that blends 86 percent wind and 14 percent landfill gas. For UC, this represented 39,000 MWh, and for CSU it represented 34,000 MWh. This commitment resulted from a two-year campaign by students and Greenpeace in which 14,000 students and faculty petitioned the universities to increase their use of renewable energy.

Since 1998, these two university systems have been aggregating their purchase of traditional electric power. Because of their very large size, they pay only 25 cents per student, or about $42,000 a year for the initial six-month contract for their aggregate green power purchase.[31]

CSU's sustainability policy, which was modeled on the UC policy, sets these goals:

- 20 percent renewable energy by 2010

- 10 MW of cost-effective, on-site renewables on campuses by 2014

- Implementation of cost-effective, sustainable building practices for planning and operations of facilities

- 15 percent reduction in energy consumption from baseline year of 2003–2004, to be evaluated in 2009–2010.[32]

ENDNOTES

1. http://www.eere.energy.gov/financing (accessed March 23, 2006).

2. http://www.nrel.gov/business_opportunities (accessed March 23, 2006).

3. Further information is available at: www.rurdev.usda.gov/rbs/farmbill/2005NOFA/nofa05navigate.html.

4. For the program's announcement, see *Federal Register* 70, no. 189 (September 30, 2005): 57250–57252. A program summary, grant application, and forms are available at: www.fpl.fs.fed.us/tmu/grant/biomass-grant.html.

5. California currently has the highest limit on net metering, at 1 MW.

6. http://epa.gov/air/oaqps/glo/designations/index.htm (accessed November 10, 2005).

   An EPA Guidance document titled "Emission Reductions from Electric-Sector Energy Efficiency and Renewable Energy Measures: State Implementation Plan Credits" was released in August 2004. This guidance allows renewable energy and energy efficiency projects to receive credit as an offset to NOx on the SIP. Montgomery County, Maryland, has received credit for its renewable purchase.

7. *Toolkit for States: Using Supplemental Environmental Projects (SEPs) to Promote Energy Efficiency (EE) and Renewable Energy (RE)* pp. 19–20.

8. www.rebuild.org

9. Barry Hilts, associate vice president, University of Pennsylvania, personal communication, August 3, 2005.

10. Walter Simpson, "Energy Sustainability and the Green Campus" *Planning for Higher Education: Journal of the Society for College and University Planning* (vol. 31 and issue no. 3): 150–58.

11. wings.buffalo.edu/ubgreen/content/policies/policy_executiveorder111_rev.html#secIIId (accessed August 25, 2005).

12. www.ithaca.edu/sustainability/news2.html (accessed October 19, 2005).

13. www.anac.org/bulletin/Fall04/bul0411-6.html#one (accessed October 19, 2005).

14. Bob Burhenn, director of energy and utilities management, Catholic University, personal correspondence, November 18, 2005.

15. www.news.harvard.edu/gazette/2005/03.10/07-renew.html (accessed November 21, 2005).

16. Sam Hummel, personal correspondence, June 24, 2005.

17. http://www.alive.pratt.duke.edu/ (accessed September 1, 2005).

18. National Association of College and University Business Officers, *College and University Business Administration*, 6th ed. (Washington, DC: NACUBO, 2000), pp. 10–14.

19. David Brixton, director, capital programs management, Arizona State University, personal communication, July 2005.

20. Ron Jackson, New Jersey Office of Clean Energy, personal communication, September 2005.

21. Tom Burke, business manager, Cerro Coso Community College, personal communication, August 2005.

22. After Pierce College began construction on the project, a voter-approved bond sale provided revenues for the college to pay cash for the project. Pierce then shortened the performance contract period to one year.

23. John F. Riley, director of purchasing and business services, Arizona State University, personal communication, July 2005.

24. The higher education institutions are participating with other California state government institutions under a third-party financing bid issued by the State Consumer Power and Financing Authority.

25. John Cusak, New Jersey Higher Education Partnership for Sustainability, personal communication, March 2006.

26. Sometimes REC forward contracts are inaccurately referred to as "futures contracts." The two are similar, but futures are very specific kinds of commodities that are bought and sold under specific conditions and strict regulations.

27. Randy Gale, director of facilities management, CSU Hayward, personal communication, December 2005.

28. NOx is a precursor to ozone.

29. Jan Hamrin and Meredith Wingate, *Regulator's Handbook on Tradable Renewable Certificates* (San Francisco: Center for Resources Solutions, 2003).

30. EPA's August 2004 guidance on NOx reductions from clean energy investments being eligible for Clean Air Act compliance is available at: www.epa.gov/ttn/oarpg/t1/memoranda/ereseerem_gd.pdf

31. Josh Lynch, Greenpeace, personal correspondence, October 5, 2005.

32. www.calstate.edu/BOT/Resolutions/sep2005.pdf (accessed October 19, 2005).

DOING THE DEAL: BUYING FROM UTILITIES AND GREEN POWER MARKETERS

This chapter addresses the purchase of renewable energy from electric utilities and renewable energy marketers and brokers. These purchases are made from renewable energy plants built and owned by a party other than the college or university. Renewable energy projects that are on or near campus are discussed in chapter 6. Some colleges and universities pursue both options, buying some green power from the local utility and installing a renewable energy project on campus. This blended approach may be the best option when an institution wants to cover a significant share of its electricity consumption with renewables, yet also have some reliable emergency backup power.

BUYING GREEN POWER AND RENEWABLE ENERGY CERTIFICATES (RECS)

Customers pay a premium for the green power products their utilities offer. This power has been generated or purchased by the utility, usually at a higher price than electricity from conventional sources. The customer doesn't actually "receive" the green power. Once the green power plant feeds its electrons into the power grid, the electrons are not differentiated from the electrons of other kinds of power plants. Instead of trying to ensure that each green buyer receives only green electrons, the utility measures the amount of electricity (in kilowatt-hours) fed into the grid by the green power generator and then assigns them to green power purchasers who have agreed to pay the premium price for them. Although green power customers do not receive any more green electrons than any other consumers, the premium they pay helps cover the extra cost of building the green power plant.

For example, if power generation in a given region costs 3 cents per kWh on the wholesale market, and the cost of generating wind power in that region is 4.5 cents per kWh, then the wind farm owner needs to find a way to cover the incremental cost of 1.5 cents per kWh. It does this by passing on the cost to customers willing to pay a green power premium.

The premium represents not only the incremental cost of green power. It also represents the environmental improvements afforded by the green power plant through reduced power and emissions output at coal-fired power plants. A green power plant can be thought of as generating two items: normal electricity and environmental benefits. The normal electricity is sold to all consumers. The environmental benefits are sold to the payers of the green premium, who thus "own" the environmental benefits, or "attributes."

The commodity being purchased with the green premium is known as a renewable energy certificate (REC). It is also known variously as a green tag, a tradable renewable certificate, and a renewable energy credit. RECs are denominated in kilowatt-hours or megawatt-hours, not in tons of emission reductions, because the calculation of exactly how much emission reduction is taking place is extremely complex and often inexact. However, the emission reductions can be estimated for each REC purchased. The estimates include reductions in carbon dioxide, sulfur dioxide, nitrogen oxides, and particulate matter. The REC also includes other positive, societal attributes of green power.[1]

RECs are sold to customers who may or may not be consumers of the electricity produced by the green power plant. In fact, while the green plant's electricity is sold to nearby customers, its RECs can be sold to customers anywhere in the country. Thus, a wind farm in Texas may supply its electricity into the local grid but have its RECs sold to customers in North Carolina.

The green power marketers play a similar role as utilities, but they are not limited to operating in a given service territory like distribution utilities.[2] While electric utilities can purchase both electricity and RECs from green power plants and sell them to their customers, the green power marketers/brokers enter into deals with green power plants to purchase only the RECs, which they then sell to interested customers either nearby or in distant locations. REC marketers also sell RECs to electric utilities, particularly in states where utilities are required to buy a portion of their electricity from renewable generators.

REC brokers, in contrast to REC marketers, generally do not take ownership of the RECs at any point; rather, they act as matchmakers between sellers and buyers. For example, the Evolution Markets Web site lists offer and bid prices for various types of RECs differentiated by geographic location, generation type, and vintage.

There are several major considerations in buying RECs:

- What renewable energy sources are represented in the RECs?
- How is the price structured?
- Are the RECs local or national?
- What is the "vintage" of the RECs?
- Will the RECs be certified, and if so, by whom?

Choosing Renewable Energy Sources

The renewable energy sources that can be bought from utilities and marketers can include wind, biomass, landfill methane, low impact hydropower, municipal solid waste (MSW), and sometimes solar energy. Usually, just one or two of these sources will be available from a given utility or marketer, but they can be blended. That is, a customer might choose between RECs that are 100 percent wind and RECs that are 50 percent wind and 50 percent landfill methane, or some other combination. The prices of the different options will vary based on the relative costs of the renewable sources.

If the local electric utility offers green power packages, an institution can simply determine which package it prefers and then sign up for it. As an alternative, it can prepare a request for proposal (RFP) for RECs and specify which sources it wants included in the REC package. Deciding on the components of the package will be an important step. The institution may decide it wants all of its RECs to be 100 percent wind. However, that may be relatively expensive, so it may decide on a package of 50 percent wind and 50 percent lower cost landfill methane. Or it may decide to lower the cost of its REC package by buying the 100 percent wind RECs but buy fewer of them. Some REC marketers have preset packages. Others are willing to work with their larger customers like colleges and universities to come up with a customized package. (See appendix C for sample green power prices.)

REC Pricing

REC prices depend on a number of factors other than the type of renewable resource, including the location of the resource, the quantity, the terms of the contract, and the vintage of the RECs (i.e., the month or year when they were generated). The price may also be affected by the level of demand for RECs in a given region and the availability of local RECs to satisfy that demand.

Green power marketers and utilities generally will offer a price break for longer-term contracts. Just as with conventionally generated electricity providers, green power marketers tend not to want to contract for too many years (generally for not more than 5 years, 10 at the most) since the market is rapidly changing and developing. Both supply and demand are increasing. Since there are a vast variety of source blends and contracts, it is difficult to generalize as to a cost analysis per kilowatt-hour. (See appendix B for sample REC prices.)

The Voluntary Market and the Compliance Market

When a college or university buys a REC, it does so voluntarily. In this case, the REC is being bought on the voluntary market. Unless mandated by its state's executive orders, higher education institutions' purchases are part of the voluntary market.

The compliance market exists in the 22 states and District of Columbia that have adopted renewable portfolio standards (RPS). An RPS requires electric utilities operating in the state to buy a certain percentage of their electricity from green power plants. The utilities can comply with this requirement by purchasing RECs in the compliance market. The same REC can be sold either on the voluntary market or on the compliance market. The distinction between the two markets is important for colleges and universities. Green power plants in RPS states will prefer to sell their RECs on the compliance market because compliance purchasers—utilities—will likely be reliable REC purchasers for multiple years, while voluntary buyers could decide to quit after a year or two. Large, long-term REC purchasers like utilities thus receive a better price break on their RECs than voluntary buyers like colleges and universities.

The compliance market is estimated to be approximately three times larger than the voluntary market. It will continue to grow as more states enact and ramp up their RPSs. Colleges and universities located in RPS states will automatically receive a portion of their electricity from renewable energy because their local utilities are required to purchase some. The cost of the RECs to the local utility will be passed on to consumers in utility rates (assuming typical regulatory rate-making practices are followed in the state). Thus, any voluntary RECs the college or university buys will be in addition to the RECs already purchased by the utility and passed on to the institution.

Local vs. National RECs

A college or university can buy RECs from either local or geographically distant renewable generators. Some colleges blend local and nonlocal or "national" sources to procure the optimal mix based on their environmental and community-related objectives and available budgets.

Local RECs

Buying local RECs helps finance local renewable energy projects, which helps the local economy. For example, wind farms provide benefits to local landowners who receive lease payments of $3,000–$5,000 per turbine.[3] Local jobs are created during the construction of renewable energy projects. Approximately 200 jobs were created during the six-month construction period of the Mountaineer Wind Energy Center in Thomas, West Virginia. American University and Catholic University purchase wind power from this facility. After construction, there are permanent jobs related to the operation and maintenance of the wind farm.

Colleges and universities in the Northwest and the Northeast are buying local, low-impact hydropower as part of their green power mixes. One of their reasons is to use their local resources to support their regional economies. The choice of local or nonlocal resources affects air quality and price. In the case of wind, prices can be lower by buying from certain areas of the country. The West has more wind resources and more and larger wind farms than the East. Hence, wind generated in the West is less expensive and national wind RECs are less costly for eastern customers than regionally generated wind RECs.

Regardless, many eastern colleges buy wind from regional wind farms. One impetus to purchase wind is to improve the local air quality. A REC purchase from a remote location does not accomplish this goal because it does not improve the local airshed by backing down emissions from coal power plants. These institutions consider it to be worth the extra money to be able to demonstrate clearly how their wind purchase affects their community's airshed. Likewise, if a state institution wishes to pursue SIP credit, remote purchases will not qualify.

CASE STUDY

Unity College—100 Percent Local Green Power

Unity College in central Maine is procuring 100 percent green power for its campus, or approximately 700,000 kWh per year. The premium is less than 1 cent per kWh, so the college is paying just under $7,000 a year. Funding for the purchase was worked into a tuition increase as there was no other available source of funds.

The resource mix is a 50:50 combination of hydropower and biomass. It was an important consideration to buy from Maine resources to support the local economy. The college buys its power through Maine Power Options, a state-run program. It considered other possible sources and made the decision based on price and the availability of local suppliers. A regional purchase made more sense than buying wind from the West.

Unity is a natural resources college where every degree is related to the environment, and the administration knows it must be increasingly sustainable. According to Mick Womersley, interim provost, the business decision to purchase green power was easy because a college "loses students when it doesn't 'walk the talk' and gains students when it does." In 2000, students began proactively pushing the administration to make the campus become more sustainable and increase its environmental initiatives. Progress has been made on many fronts, including the decision to purchase green power.

Student retention has increased during a time when the demographics suggested a decrease. The college has received positive press on its green power purchase, including radio and TV coverage.[4]

National RECs

In many cases, national RECs will be less expensive than local ones.[5] An eastern college or university contemplating the purchase of RECs must decide whether to buy the cheapest RECs or support local economic development and improvements to the local airshed. National RECs have another advantage. Because there are more national RECs than local ones, REC buyers have more options to choose from when considering a REC purchase. They have more choices regarding the type of renewable energy and the vintage of the RECs.

Some areas do not have any local renewable generation, so green buyers have no choice but to buy national RECs. For example, there are presently no wind farms in the Southeast. For this reason, Nicholas School of the Environment and Earth Sciences at Duke University chose to buy wind power RECs from Kansas.

REC Vintages

Another consideration in buying RECs is the "vintage" or age of the REC. While most REC buyers would expect their RECs to have been generated in the current year, it is possible to purchase RECs from previous years. These certificates represent renewable energy that was generated, but its RECs were never sold. Older vintage RECs tend to be less expensive than current-year RECs. If the RECs are too old, they will not be certified by independent REC certification bodies. The matter of vintage may be important to a one-time buyer. But to consumers that make a commitment to renewable energy and plan to buy them on an ongoing basis, the matter is probably irrelevant in most cases.

New and Old Sources

Many utilities have included green power in their energy mixes for many years. A prime example is large hydropower companies in the Northwest and Northeast. If a college or university is simply buying renewable energy that has been part of its utilities' mix, it is considered an "old" source. Many observers do not view a purchase of old renewables as particularly relevant since it is not pushing the renewable energy market forward or stimulating new demand. New sources are defined by the date that the renewable energy system went online.

According to the EPA Green Power Partnership, "new" renewables refer to renewable facilities that have been developed specifically to serve the green power market. Generally, a new renewable resource is one that began generating after January 1, 1997.[6] Other organizations have slightly different definitions. As time passes, the date for what is considered new is moved up by the certification bodies, so that older RECs no longer can be certified.

Certification

RECs can be certified by independent entities to ensure that suppliers' claims are accurate and that the RECs meet certain quality standards. Certification also ensures that the RECs are not being used for compliance with a mandate or for another reason.[7] Certification is a process by which various criteria for renewable energy are verified. The two nonprofit organizations that certify green power are the Center for Resource Solutions (CRS) and the Environmental Resources Trust (ERT). The major certifications are CRS's Green-e and ERT's EcoPower. REC marketers must pay for certification and that cost is passed on to the buyers. If an institution wants to buy certified green power, there are minimum percentage requirements for new sources. There are differing opinions on the necessity of certification. Some view certification as an unnecessary added expense.

Green-e

Green-e has certified 56 REC products and has over 120 participating marketers as of early 2006. CRS defines "new" as clean energy from a facility that began operations after January 1, 1999. Green-e must be a fully aggregated REC and contain all the environmental attributes of the unit of renewable generation. It cannot be used to also meet other state or local requirements such as RPSs. This is to avoid the issue of double counting—when two claims are made on the same ton of reductions. For example, a utility is double counting if it uses green power to satisfy its RPS requirements and also sells the environmental attributes as RECs to another entity.

Green-e RECs must also be generated in the calendar year, the first three months of the next calendar year, or the last six months of the prior calendar year. To be certified, 50 or 100 percent electricity is required to be supplied from one or more eligible renewable resources. [8]

Environmental Resources Trust, Inc. (ERT)

ERT certifies the environmental attributes of clean energy that meet its requirements for EcoPower. The requirements are outlined in the *Uniform National Standard for EcoPower Renewable Energy Certificates*. The renewable energy source must be 50 percent "new" for certification. As of January 1, 2007, this will be increased to 100 percent. "New" is defined as a source that has come online on or after January 1, 1998.[9] EcoPower RECs should include the indirect environmental impacts but are not required to provide direct emission reduction credit. The renewable resource must be connected to the grid. This distinction is important for ERT to avoid double counting. By ERT definitions:

- *Indirect reductions* of air and water emissions are caused when renewable energy sources displace demand for electricity generated from other, more polluting sources, such as fossil fuel combustion.

- *Direct reductions* of greenhouse gas emissions result when a powerful greenhouse gas is converted on site by the renewable energy generator into a less potent greenhouse gas (e.g., the generation of electricity through the combustion of landfill gas methane to carbon dioxide).[10]

Both CRS and ERT require that RECs have expiration dates and vintages. Both certifications consider such certifiable renewable energy sources as wind energy, solar energy, some biomass fuel sources, low-impact hydropower, and geothermal. EcoPower RECs also include as eligible renewable energy sources tidal and wave energy and hydrogen fuels derived from other eligible renewable energy sources.

Each organization has certain requirements for the feedstock for the biomass in order to be certifiable. For example, Green-e accepts wood waste, agricultural waste, animal waste, and landfill gas. It does not certify municipal solid waste (MSW).

▶ TRACKING SYSTEMS FOR RECS

REC tracking systems verify compliance with renewable energy requirements, substantiate marketing claims, and protect against trading abuses and misrepresentation.[11] Electronic certificate tracking systems exist in New England (New England Power Pool), Texas (operated by the Electric Reliability Council of Texas), and Wisconsin (operated by the Wisconsin Public Service Commission).[12]

In the PJM grid, the Generation Attribute Tracking System (GATS), was launched in 2005. GATS is an environmental and emissions attributes tracking system for electric generation. PJM is the electric grid that includes but is not limited to service in Pennsylvania, New Jersey, Maryland, parts of western Virginia, Washington, D.C., and Delaware. This system will provide data for the information disclosure requirements for renewable portfolio standards.

Western Renewable Energy Generation Information System (WREGIS) is a new system that tracks and verifies the generation of clean energy and also creates RECs. It is a voluntary, independent system to account for wholesale renewable energy transactions. The Western Governors Association, the California Energy Commission, and the Western Regional Air Partnership are the sponsors of WREGIS, which became operational at the end of 2005. Fourteen states and two provinces are participating.[13]

New Jersey has a new Solar Renewable Energy Certificates Program, which is a tracking and trading system for solar renewable energy certificates (SRECs).[14] This system is the first of its kind in the United States. Each SREC represents 1 MWh of solar production. Other states are likewise considering the establishment of their own REC certification systems.

▶ THE PURCHASE PROCESS

After deciding to purchase RECs or green electricity and getting approval for the purchase, the college or university will need to ask the following questions:

- Who will develop, administer, and judge RFP—internal staff or energy consultants?
- How much funding is available for the purchase?
- How will the purchase be funded?
- What percentage of total electricity use will be "supplied" by the RECs ?
- Are there possible partners for an aggregate purchase?
- Are there additional funding sources (i.e., state grants, SEP funding)?
- Is there a way to work with the state to obtain SIP credit?

Finding and Comparing Marketers

Many utilities offer green power options to their customers. A first step is to contact the local utility to determine if it has a green power product. Green-e has a list of marketers whose products are certified. For on-site generation, the equipment marketers and installation companies can be located through industry associations. The Department of Energy's Green Power Network also has information on finding and comparing marketers at: www.eere.energy.gov/greenpower/buying/buying_power.shtml.

An excellent source of information and tools is EPA's Green Power Partnership (www.epa.gov/greenpower), which has 500-plus partners, including more than 40 higher education institutions. Partners make a commitment to purchase a certain amount of green power based on overall consumption. Typically, larger partners purchase more clean energy and smaller partners buy a higher percentage of green power. The partnership has a Green Power Leadership program and a prestigious annual awards program. Concordia University, California State University at Hayward, College of the Atlantic, Loyola Marymount University, Pennsylvania State University, Carnegie Mellon University, University of Pennsylvania, and University of Colorado, Boulder, are some previous Green Power award winners. There are no fees for joining the Green Power Partnership or using its tools.

Questions to ask when comparing marketers include:

- What are the types of green power options?

- What is the mix of sources?

- Are blends of sources available?

- What is the price differential for various sources, products, and contract lengths?

- Is it from a local or national source?

- What are the length and terms of the contract?

- Are fixed price contracts or contracts for differences available?

- Is the product a REC or electricity through the grid?

- Is the green power certified and if so, by what organization?

- Is an aggregate purchasing plan available?

The Contract and Procurement Process

Utilities with green power programs have different methods of structuring the agreement. Often, utilities and green power marketers purchase clean energy at a premium charge on top of standard charges. Municipal utilities tend to have more of a community focus. The Sacramento Municipal Utility District (SMUD), for example, matches 40 percent of a customer's premium to help develop new renewable power plants. Other utilities charge fixed percentages of a customer's load for clean energy purchases. In others, the end user can determine which percentage is optimal.

There are a few typical price protection methods. For example, green power customers do not incur the fuel-cost adjustments in some utilities programs, in which customers are prorated based on the percentage of renewable energy purchased. Second, some utilities offer fixed-price green power products. Also, some utilities exempt green power customers from fees associated with environmental costs from fossil fuel generation, air quality riders, and/or conventional power surcharges. As a result of these methods, the effective premiums of green purchases have been reduced significantly over the last few years.[15]

Traditionally, marketers prefer longer contracts and will offer a lower price for longer terms. However, some marketers of renewable energy will not go beyond a few years at this point, since the market is developing so quickly. College of the Atlantic had planned to sign a 20-year contract, but because of a change of marketer was only able to sign a 10-year contract. Buyers often prefer a ramping-up period. The University of Pennsylvania signed a 10-year wind contract after an original contract of five years. Other examples of institutions which have ramped up include University of Buffalo, SUNY, and the University of Colorado, Boulder. A combination of long- and short-term contracts may be most advantageous after an initial one- to two-year test period.

If a state has a significant RPS, some marketers' prices are increasing because the supply is tight. However, since a federal production credit was included as part of the Energy Policy Act of 2005, the credit is extended through December 2007. More renewable energy supply is likely to come online as a result.

▶ BUYING RECS AS A PRICE HEDGE

One advantage of renewable energy sources like solar and wind is that they have no fuel costs and thus can be counted on to generate electricity at a stable price over time. However, utilities and other green power marketers do not typically pass on this price stability to their customers. Instead, the marketers simply charge a premium for green power in addition to whatever utility bill the customer already pays. That is, even if a college or university purchases wind RECs equivalent to 100 percent of its electricity consumption, its electric bills will nevertheless increase like everyone else's as the price of conventional fuels rise.

There are now ways to purchase a type of REC that serves as a hedge against the volatility of fossil fuel prices. These RECs are structured in such a way that if the college or university enters into a long-term fixed-price contract for, say, 100 MWh worth of RECs per year, then those 100 MWh of its annual electricity usage will not be subject to fossil fuel price volatility. If the college or university enters into a contract to purchase 100 percent of its estimated future electricity consumption in this manner, it will completely insulate itself from fossil fuel price shocks and volatility for the duration of the contract.

To date, this approach has been used mainly for wind power. Figure 5-1 is an example of a wind hedge compared to a normal electric bill on a price per kilowatt-hour basis. The wind hedge gives the institution a stable price for electricity. It will cost more than conventional electricity at first, but over time, as conventional fuel prices increase, the wind hedge is a better deal. If fuel prices decrease, the college or university ends up losing money. Thus, there is some risk with wind hedges, although most buyers think it is a good gamble to bet on prices increasing rather than decreasing. Hedge products usually have terms of 10 years or longer.

FIGURE 5-1: USING WIND RECS TO HEDGE FUEL PRICE VOLATILITY

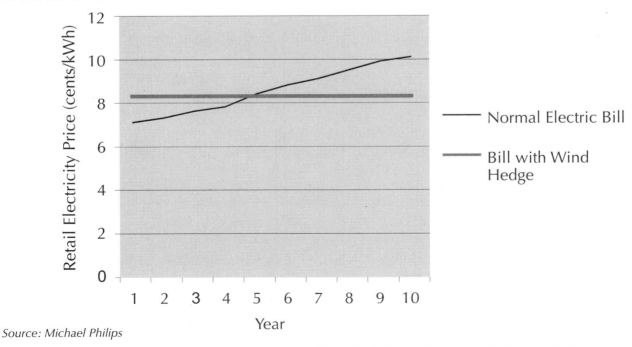

Source: Michael Philips

CASE STUDY

Concordia University—Fixed-Price Wind Hedge

Concordia University in Austin, Texas, was the first higher education institution in the country to buy 100 percent renewable power for its campus. The blend is 80 percent wind, 17 percent biogas, 2 percent hydropower, and 1 percent solar. When Concordia purchased its fixed-price contract, Austin Energy Green Choice had set the price of the wind hedge slightly higher than its average retail electricity price because it cost slightly more to acquire wind-generated electricity. Therefore, Concordia started out paying more for its wind power than it would have for conventional electricity.

The university's contract guarantees the green power fuel price for 10 years at 2.85 cents per kWh. With spiking gas prices, this fuel charge remains constant. Over time, the price of fossil fuels, especially natural gas, has increased, and Austin Energy's average retail price has increased beyond the price of Concordia's wind hedge. According to facilities manager Ron Petty, there have been three price increases in conventional electricity since the university signed the wind contract.

Concordia is now making money every year in the form of lower utility bills. The charge for a traditional contract beginning in January 2006 will be 3.63 cents per kWh, a difference of 0.78 cents per kWh. Based on 2005 usage and similar weather, operational, and occupancy patterns, Concordia would have to spend $50,600 more in 2006 if it were not in this contract.

The budget predictability is a very appealing aspect of the project. Concordia can now plan without concern about electricity price increases because the energy numbers are solid until 2011. [16]

The president, faculty, staff, and students at Concordia view a green power purchase as part of its mission, which includes the stewardship of the earth. Also, the university has done energy and water management projects to improve the efficiency of the campus.

Among the utilities that offer wind hedges is Austin Energy, the municipally owned utility in Austin, Texas. Austin Energy entered into its own long-term fixed-price purchase contract with local wind farms. With a fixed-price contract, an institution can at least be certain about that part of its energy budget. In some areas, wind power now costs less than traditionally generated power for customers with fixed-price contracts.

Only an electric utility like Austin Energy can guarantee that a college or university's wind hedge will keep its electricity bills constant, at least for those megawatt-hours of consumption covered in the hedge contract. [17] If an institution decides to go 100 percent wind, then the utility guarantees that its utility bill (per kilowatt-hour) will not increase at all for the duration of the hedge contract.

Conversely, non-utility REC suppliers have no control over utility bills and so cannot guarantee a fixed price for electricity. However, what they can do is reimburse buyers for electricity price increases by offering a financial hedge against price increases in the form of a "contract for differences." This is an instrument that is already in use by some colleges and universities for the purchase of conventional electricity.

▶ CONTRACT FOR DIFFERENCES

A contract for differences is a purely financial forward contract (as opposed to a contract to buy electricity) in which a college or university can insulate itself from the volatility of conventional electricity prices for a contract period of 5 to 10 years or longer. Also known as a fixed-for-floating swap, a contract for differences (CFD) involves a college or university entering into an agreement with an electricity supplier wherein the two parties pay each other as conventional electricity prices fluctuate.

CFDs can be entered into without involving renewable energy. They simply hedge the price of conventional electricity. Under a green hedge, the college or university can receive RECs as part of the deal. In such a case, the supplier would be a green power provider such as a wind power marketer or broker.

All CFDs are based on the wholesale price of electricity. To enter into a CFD with a green electricity supplier, the two parties agree on a "strike" price for electricity. The strike price is a wholesale price and is the reference point for determining which party pays the other as wholesale electricity prices fluctuate. It is often set at a premium above the current wholesale price. Once the parties sign the contract, any time the spot wholesale price goes higher than the strike price, the supplier pays the college or university the difference between the two prices; and any time the spot wholesale price drops below the strike price, the college or university pays the supplier the difference.

For example, assume that a college or university is paying 7.5 cents/kWh for its retail electricity. Retail prices will tend to track wholesale prices. Assume that the wholesale price is 6 cents/kWh.[18] The green electricity supplier and the college or university might agree to a strike price of 7 cents/kWh. They also agree that the CFD will apply to a certain number kWhs over a certain number of years. Those kWhs will represent the RECs the college or university will take title to, although the price of those RECs will vary as wholesale electricity prices vary in relation to the strike price. If at some point the spot wholesale electricity price rises to say, 8.0 cents/kWh, the renewable energy supplier will reimburse the college or university at the rate of 1.0 cent/kWh (the strike price of 7.0 cents minus 8.0 cents). In the event that electricity prices decrease to say, 6 cents/kWh, the college or university will pay the supplier 1.0 cent/kWh (7.0 cents minus 6 cents).[19]

In the example in figure 5-2, the college agrees to a strike price of $50 per mega-watt-hour (or 5 cents per kWh) for wind power. When conventional electricity exceeds that price, the wind marketer pays the college within the darkly shaded areas above the 5-cent line. When conventional electricity drops below 5 cents, the college pays the marketer within the lightly shaded area.

Although the college is exposed to a liability if conventional electricity prices drop, it is saving money because of the lower electricity price. The contract is structured so that in the event of an electricity price decrease, the college pays the renewable energy marketer out of the savings on its lowered utility bill.

FIGURE 5-2: SCHEMATIC OF WIND POWER CONTRACT FOR DIFFERENCES SET AT 5C/KWH

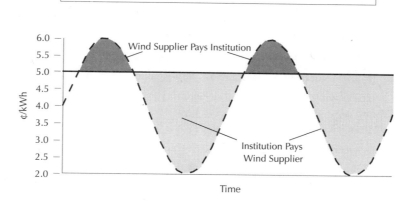

Source: New York State Energy Research and Development Authority, "Using Wind Power to Hedge Volatile Electricity Prices for Commercial and Industrial Customers in New York," May 2003

▶ CALCULATING THE ENVIRONMENTAL BENEFITS OF RECS

Many colleges and universities are interested in quantifying the environmental benefits of green power purchases. Generally, a marketer will provide the calculations as part of its proposal. Common metrics include:

- Emissions prevented in pounds per year for carbon dioxide, sulfur dioxide, nitrogen oxides, and mercury

- The carbon dioxide equivalent of planting x number of trees

- The equivalent of x number of miles not driven

- The equivalent of x number of cars taken off the road

The National Renewable Energy Laboratory (NREL) of the Department of Energy has a Renewable Energy Converter that quantifies displaced emissions that result from a renewable electricity purchase.[20] The calculator is set up for biomass, concentrating solar, geothermal, photovoltaics, and wind. It converts to metrics that include:

- Annual household electricity needs met

- Tons of carbon monoxide, coal, nitrogen oxides, and sulfur dioxide reduced

- Cubic feet of natural gas displaced

Emission factors for various power plants are reported to the EPA eGRID, a searchable database on air quality. eGRID also shows generation of electricity by fuel type and the fuel mixes in different regions.[21]

EPA's Power Profiler lets the user know how a particular region's fuel mix compares to the national averages; compares emission rates of the area to national averages, and calculates the emissions that can be attributed to a business or facility based on zip codes.[22]

ENDNOTES

1. www.epa.gov/greenpower/whatis/glossary.htm.

2. A list of green power suppliers and brokers is available at: www.eere.energy.gov/greenpower/markets/marketing.shtml?page=2.

3. www.awea.org/faq/tutorial/wwt_economy.html#In percent20what percent20other percent20ways percent20does percent20wind percent20energy percent20benefit percent20the percent20economy (accessed October 4, 2005).

4. Mick Womersley, interim provost, Unity College, personal communication, July 28, 2005.

5. For example, stronger prevailing wind in the western United States means it is less expensive to generate wind in the West than in the East. Thus, wind RECs from the West are generally less expensive than wind RECs from the East.

6. www.epa.gov/greenpower/whatis/glossary.htm (accessed July 17, 2005).

7. E. A. Holt and L. Bird, *Emerging Markets for Renewable Energy Certificates: Opportunities and Challenges* (Golden, CO: National Renewable Energy Laboratory, 2005), p. 17.

8. http://green-e.org (accessed August 3, 2005).

9. Uniform National Standard for EcoPower Renewable Energy Certificates, pp. 3–6, available at: www.ert.net/pubs/EcoPowerStandard.pdf (accessed February 10, 2006).

10. Ibid., p. 4.

11. Holt and Bird, *Emerging Markets for Renewable Energy Certificates*, p. 15.

12. Ibid., p. 16.

13. www.westgov.org/wga_wregis.htm (accessed November 7, 2005).

14. www.njcep.com/srec/ (accessed November 16, 2005).

15. Lori Bird, "Fuel Price Stability Benefits of Renewables: Adding Value for Green Power Customers," presentation at Eighth National Green Power Marketing Conference, November 2003, available at: www.eere.energy.gov/greenpower/conference/8gpmc03/bird03.pdf (accessed October 6, 2005).

16. Ron Petty, facilities manager, Concordia University, September 15, 2005, personal correspondence.

17. Other utilities that are either offering or considering fixed wind power prices to hedge conventional energy prices are Xcel in Colorado, Oklahoma Gas & Electric, and Portland General Electric in Oregon.

18. The 1.5 cent spread between retail and wholesale is based on load shape, capacity reserves, ancillary services, and losses.

19. It should be noted that while the typical electricity bill consists of charges for electricity generation, transmission and distribution, a CFD only hedges the generation component of the total electricity price.

20. http://analysis.nrel.gov/databook/convert.asp (accessed September 26, 2005).

21. www.epa.gov/cleanenergy/egrid/index.htm (accessed November 7, 2005).

22. www.epa.gov/cleanenergy/powerprofiler.htm (accessed November 7, 2005).

DOING THE DEAL: ON-SITE GENERATION

I n some ways, planning and undertaking a renewable energy project is like any other capital project because it involves design work, costing out of materials and labor, getting permits, and so forth. But in other ways, a renewable energy project is different. The college or university is building a small power plant, so there are electric utility interconnection issues. There is also a revenue stream from the project—in the form of reduced electric bills or payments from the utility for the electricity—that will affect decisions on project sizing and cash flow calculations. There may also be ownership issues. Will the college or university own the project, will an energy service provider own it, or will it be a community project with co-owners?

This chapter addresses the major steps and considerations in planning and building a renewable energy project on campus or near the campus: renewable energy resource assessment and site selection; initial economic analysis; ownership options; interconnection issues; and design and preconstruction. The steps are not necessarily sequential, and in fact some may need to be undertaken simultaneously. Others may not be relevant in a given situation or for a given technology.

▶ RESOURCE ASSESSMENT AND SITE EVALUATION

Unless a college or university is certain which renewable technology it wants to pursue, the first step will be to assess the availability of all renewable energy resources located on or near the campus. This initial assessment should seek to identify each renewable energy resource and quantify its long-term availability. Extensive information is available online, especially from the National Renewable Energy Laboratory but also from state energy agencies and other organizations. These general resources show, for example, prevailing wind conditions in a given region of a state. Data for some renewable energy sources, such as wind and biomass, must be site specific. The assessment will be easiest for solar energy, easy but time-consuming for wind, and most complex for biomass given the variety of biomass feedstock fuels, which range from agricultural products, to landfill methane, to animal wastes.

Wind Assessment

A wind resource assessment will determine the best site (or sites) for locating a wind project. It can be based on production numbers from nearby wind turbines or on public data from nearby meteorological towers. If there are no such towers or if the data are unavailable, the college or university can hire a meteorologist to prepare a production estimate. An expensive and time-consuming option is to rent a tower with wind data collection equipment (anemometers) for a year, at a cost of approximately $8,000, including installation and removal. More than one tower may be needed if there is more than one potential location for the wind turbine.

It's a good chance that the available locations on a campus are not the best wind sites in the larger area. Nearby properties may have better wind conditions and/or proximity to transmission lines (for interconnection with the electric utility). The lease payment the university would make to another landowner will be a small part of the total project costs, and will be more than offset by the value of the increased kWh output from the superior site. As both Iowa Lakes Community College and Carlton College found in preparing their wind power projects, the wind resource was better one mile from campus than directly on campus.

Solar Assessment

The amount of solar radiation striking the campus will not vary from location to location, so a solar radiation assessment will not be needed to make a project siting decision. However, determining the average annual amount of solar radiation will be useful in evaluating the economic feasibility of the project. The National Renewable Energy Laboratory's online tools help assess the solar radiation at any geographic location in the United States.[1] Its PVWATTS software provides easy-to-understand performance estimates for grid-connected photovoltaic systems.

Selecting a suitable site for a solar project can be as simple as determining which unshaded campus rooftops are preferable (and are able to withstand the extra weight.) However, Pierce College found that its preferred location on the gymnasium roof would violate local building codes and so had to select an alternative site over a parking lot.

Biomass Assessment

A biomass assessment should initially consider the full range of biomass resources. Although the availability and cost of the resources will vary by location, the most widely available resources are generally landfill methane, urban wood waste, and forest residues. In rural areas, crop residues and animal wastes should be considered, although odor issues can arise with the animal wastes. Crops grown specifically for fuel will tend to be cost-prohibitive in most instances, and can be dropped from most resource assessments.

Resource assessments can also consider organic waste products from certain industries. (See section on Community Ownership for a list of common industries that may have useful waste streams.)

As a general rule, the assessment should consider resources within a 50-mile radius of the campus; otherwise, transportation costs become an issue. Some biomass projects use smaller radii. For landfill projects that transport their fuel via pipeline, there is no such rule of thumb. Hudson Valley Community College transports its landfill gas less than a mile, while UCLA has a 4.5-mile pipeline.

▶ INITIAL ECONOMIC ANALYSIS

The data from the resource assessment and site evaluation are used to calculate the cost per installed kW of each renewable technology. A careful analysis of the data will be essential to the decision making, but for an initial economic analysis, the general technology costs described in chapter 3 will suffice—$5,500–$6,300 per installed kilowatt for solar PV; $1,000–$1,500 for wind; and $1,300–$3,000 for biomass (lower if biomass will be used for space heating and not electricity generation). These prices assume that no batteries or other storage system will be included.

College- or University-Owned Projects

An economic analysis of projects that will be built and owned by the college or university should include the following variables:

- Base year avoided retail tariff
- Estimated project size (installed capacity in kilowatts) and annual electricity production (in kilowatt-hours), determined by out-of-pocket costs (including cost of batteries), physical area of proposed site (e.g., size of roof), net metering limits, utility negotiations, ownership structure selected, and load for critical functions (if the project will provide power during an emergency)
- Capital cost
- Capital cost after all available federal, state, or utility grants and rebates
- Down payment, if any
- Estimated annual financing payment
- Estimated annual operations and maintenance (O&M) costs. An O&M service contract is recommended for smaller installations; facilities departments usually handle O&M for smaller installations, especially PV systems.
- Estimated annual fuel cost (for biomass projects)
- Estimated annual savings from avoided retail electricity bills (initial calculation can assume all renewable electricity will be consumed on campus)
- Estimated income from sale of RECs and other environmental benefits (e.g., NOx SIP credits, carbon credits)
- Estimated annual change in renewable system production (supplied by bidders)
- Annual escalation factor for retail tariff changes (supplied by state utility commission or energy office)
- Annual escalation factor (if any) for renewable energy project bid price changes (supplied by bidders)
- Discount rate (for calculating NPV)
- Financing term (up to 20 years)
- Lifetime of the project (20–30 years)

Later in the planning and budgeting process, a more accurate analysis will incorporate cost figures provided by equipment and service vendors.

Vendor-Owned Projects

An economic analysis of projects that will be undertaken under a performance contract, lease, or other third-party service model should consider these factors:

- Base year avoided retail tariff
- Estimated annual savings from avoided retail electricity bills
- Estimated income from sale of RECs and any other environmental benefits (e.g., NOx SIP credits, carbon credits)
- Down payment, if any
- Annual performance contract or lease payment. If using third-party service model, cost per kilowatt-hour for purchased renewable electricity, estimated output of project, and estimated annual cost of the renewable kilowatt-hour
- Estimated annual change in renewable system production (supplied by bidders)
- Annual escalation factor for retail tariff changes (supplied by state utility commission or energy office)
- Annual escalation factor (if any) for renewable energy project bid price changes (supplied by bidders)
- Discount rate (for calculating NPV)
- Financing term (up to 20 years)
- Lifetime of the project (20–30 years)

▶ CASH FLOW ANALYSIS

Many variables affect the costs and revenues of a renewable energy project, and every college or university will have site-specific issues that affect the cash flow analysis. Figures 6-1, 6-2, and 6-3 use some reasonable assumptions to illustrate that renewable energy investments can be economically viable. These spreadsheets do not include preconstruction and installation costs. They assume commercial interest rates, normal operations and maintenance costs, and the availability of state renewable energy incentives. The solar example (figure 6-3) assumes that the solar electricity will offset a high utility electricity rate.

Figure 6-1 is a sample cash flow analysis for a 750 kW wind turbine operating in good wind conditions and generating just under 4.7 MWh per year. The analysis assumes that the project will receive revenues from power sales to the local utility and from a state incentive payment. It assumes no income from the sale of RECs or from any tax benefits. The wind production tax credit and depreciation are shown for colleges and universities that may want to consider soliciting private equity investors attracted to the tax advantages. The financing terms of 7 percent interest for 10 years are conservative; in 2006 it is possible to obtain 5 percent for 15 years. A standard warranty is assumed for the first 10 years of operation, with an extended warranty for five years after that. The analysis shows that the turbine will generate a net positive cash flow of $38,781 per year.

FIGURE 6-1: SAMPLE CASH FLOW ANALYSIS FOR A 750 KW WIND TURBINE (LARGE)

	Year 1	Year 2	Year 3	Year 4	Year 5	Year 6	Year 7	Year 8	Year 9	Year 10	Year 11	Year 12	Year 13	Year 14	Year 15	Total
Revenues																
on 4,700,000 kWh/yr																
PPA @ $.035/kWh	$164,500	$164,500	$164,500	$164,500	$164,500	$164,500	$164,500	$164,500	$164,500	$164,500	$164,500	$164,500	$164,500	$164,500	$164,500	
State prod pmt @ $.015/kWh	$70,228	$70,228	$70,228	$70,228	$70,228	$70,228	$70,228	$70,228	$70,228	$70,228						
Total Revenues	$234,728	$234,728	$234,728	$234,728	$234,728	$234,728	$234,728	$234,728	$234,728	$234,728	$164,500	$164,500	$164,500	$164,500	$164,500	$3,169,780
Expenses																
Mgmt fee	$22,500	$22,500	$22,500	$22,500	$22,500	$22,500	$22,500	$22,500	$22,500	$22,500	$22,500	$22,500	$22,500	$22,500	$22,500	$337,500
Service warr	$22,000	$22,000	$22,000	$22,000	$22,000	$23,000	$23,000	$23,000	$23,000	$23,000	$35,000	$35,000	$35,000	$35,000	$35,000	$400,000
Elec usage	$1,000	$1,000	$1,000	$1,000	$1,000	$1,000	$1,000	$1,000	$1,000	$1,000	$1,000	$1,000	$1,000	$1,000	$1,000	$15,000
Land rent	$4,000	$4,000	$4,000	$4,000	$4,000	$4,000	$4,000	$4,000	$4,000	$4,000	$4,000	$4,000	$4,000	$4,000	$4,000	$60,000
Insurance	$10,000	$10,000	$10,000	$10,000	$10,000	$11,000	$11,000	$11,000	$11,000	$11,000	$13,000	$13,000	$13,000	$13,000	$13,000	$170,000
Total Expenses	$59,500	$59,500	$59,500	$59,500	$59,500	$61,500	$61,500	$61,500	$61,500	$61,500	$75,500	$75,500	$75,500	$75,500	$75,500	$982,500
Operating Cash	$175,228	$175,228	$175,228	$175,228	$175,228	$173,228	$173,228	$173,228	$173,228	$173,228	$89,000	$89,000	$89,000	$89,000	$89,000	$2,187,280
Debt Service	$136,447	$136,447	$136,447	$136,447	$136,447	$136,447	$136,447	$136,447	$136,447	$136,447	-	-	-	-	-	$1,364,470
Coverage ratio	1.21	1.21	1.21	1.21	1.21	1.2	1.2	1.2	1.2	1.2						
Project Reserves	$38,781	$38,781	$38,781	$38,781	$38,781	$36,781	$36,781	$36,781	$36,781	$36,781	$89,000	$89,000	$89,000	$89,000	$89,000	$822,810
Fed PTC value	$79,591	$81,183	$82,807	$84,463	$86,152	$87,875	$89,633	$91,425	$93,254	$95,119						$871,502
Dep. Value	$105,876	$64,167	$64,167	$54,542	$32,084											$320,836
Total tax value																$1,192,338

Assumptions

Project cost $1,283,344
Debt: $958,344 @ 7% int., 10 yr term
Equity: $325,000 recovered in form of Federal PTC and depreciation
Federal PTC escalation @ 2%/yr
ACRS depreciation 5 yrs @ 25% tax bracket

Source: Prepared by Marty Silber, Consultant, Washington, DC

FIGURE 6-2: SAMPLE CASH FLOW ANALYSIS FOR A 10 KW WIND TURBINE (SMALL)

	Year 1	Year 2	Year 3	Year 4	Year 5	Year 6	Year 7	Year 8	Year 9	Year 10	Year 11	Year 12	Year 13	Year 14	Year 15
Revenues															
Electricity Sales	$4,320.00	$4,428.00	$4,538.70	$4,652.17	$4,768.47	$4,887.68	$5,009.88	$5,135.12	$5,263.50	$5,395.09	$5,529.97	$5,668.21	$5,809.92	$5,955.17	$6,104.05
Total Revenues	$4,320.00	$4,428.00	$4,538.70	$4,652.17	$4,768.47	$4,887.68	$5,009.88	$5,135.12	$5,263.50	$5,395.09	$5,529.97	$5,668.21	$5,809.92	$5,955.17	$6,104.05
Expenses															
O&M (1 percent of cost per annum)	$349.50	$349.50	$349.50	$349.50	$349.50	$349.50	$349.50	$349.50	$349.50	$349.50	$349.50	$349.50	$349.50	$349.50	$349.50
Total Expenses	$349.50	$349.50	$349.50	$349.50	$349.50	$349.50	$349.50	$349.50	$349.50	$349.50	$349.50	$349.50	$349.50	$349.50	$349.50
Operating Cash Flow	$3,970.50	$4,078.50	$4,189.20	$4,302.67	$4,418.97	$4,538.18	$4,660.38	$4,785.62	$4,914.00	$5,045.59	$5,180.47	$5,318.71	$5,460.42	$5,605.67	$5,754.55
Debt Service 15 Years; 5 percent	$3,367.16	$3,367.16	$3,367.16	$3,367.16	$3,367.16	$3,367.16	$3,367.16	$3,367.16	$3,367.16	$3,367.16	$3,367.16	$3,367.16	$3,367.16	$3,367.16	$3,367.16
Net Cash Flow	$603.34	$711.34	$822.04	$935.50	$1,051.81	$1,171.02	$1,293.21	$1,418.46	$1,546.84	$1,678.43	$1,813.30	$1,951.55	$2,093.26	$2,238.50	$2,387.38
Coverage Ratio	1.18	1.21	1.24	1.28	1.31	1.35	1.38	1.42	1.46	1.50	1.54	1.58	1.62	1.66	1.71
Electricity Rate	0.180	0.185	0.189	0.194	0.199	0.204	0.209	0.214	0.219	0.225	0.230	0.236	0.242	0.248	0.254

Assumptions
Size: 10 KW Small Wind-Non Grid Tied System   Inputs / Watt
System: Bergey 10 KW Excel-S with a Gridtek Power Processor
# of Watts: 10 KW
Cost: 34,950.00
Total kW Capacity: 10
Daily Power Generation: 65.75 kWh
Annual 24000 kWh

Source: Prepared by Marty Silber, Consultant, Washington, DC

FIGURE 6-3: SAMPLE CASH FLOW ANALYSIS FOR A 100 KW SOLAR PHOTOVOLTAIC INSTALLATION

	Year 1	Year 2	Year 3	Year 4	Year 5	Year 6	Year 7	Year 8	Year 9	Year 10	Year 11	Year 12	Year 13	Year 14	Year 15
Revenues															
Rec Sales .75 cents per kWH	$1,450.88	$1,450.88	$1,450.88	$1,450.88	$1,450.88	$1,450.88	$1,450.88	$1,450.88	$1,450.88	$1,450.88	$1,450.88	$1,450.88	$1,450.88	$1,450.88	$1,450.88
Electricity Sales $49,201.16	$34,821.00	$35,691.53	$36,583.81	$37,498.41	$38,435.87	$39,396.77	$40,381.68	$41,391.23	$42,426.01	$43,486.66	$44,573.82	$45,688.17	$46,830.37	$48,001.13	
Total Revenues $50,652.04	$36,271.88	$37,142.40	$38,034.69	$38,949.28	$39,886.74	$40,847.64	$41,832.56	$42,842.10	$43,876.88	$44,937.53	$46,024.70	$47,139.04	$48,281.25	$49,452.01	
Expenses															
O&M 1 percent of cost per annum	$3,250.00	$3,250.00	$3,250.00	$3,250.00	$3,250.00	$3,250.00	$3,250.00	$3,250.00	$3,250.00	$3,250.00	$3,250.00	$3,250.00	$3,250.00	$3,250.00	$3,250.00
Total Expenses	3,250.00	3,250.00	3,250.00	3,250.00	3,250.00	3,250.00	3,250.00	3,250.00	3,250.00	3,250.00	3,250.00	3,250.00	3,250.00	3,250.00	3,250.00
Operating Cash Flow $47,402.04	$33,021.88	$33,892.40	$34,784.69	$35,699.28	$36,636.74	$37,597.64	$38,582.56	$39,592.10	$40,626.88	$41,687.53	$42,774.70	$43,889.04	$45,031.25	$46,202.01	
Debt Service ($31,311.24) 15 Years; 5 percent	($31,311.24)	($31,311.24)	($31,311.24)	($31,311.24)	($31,311.24)	($31,311.24)	($31,311.24)	($31,311.24)	($31,311.24)	($31,311.24)	($31,311.24)	($31,311.24)	($31,311.24)	($31,311.24)	
Coverage Ratio	($1.05)	($1.08)	($1.11)	($1.14)	($1.17)	($1.20)	($1.23)	($1.26)	($1.30)	($1.33)	($1.37)	($1.40)	($1.44)	($1.48)	($1.51)

Assumptions

Size: 100 KW PV system		
	Inputs / Watt	
Installed Cost	6.50	Per watt
Rebates	3.25	Per watt
Net PV Cost	3.25	
# of Watts	100000	
Cost	325,000.00	
Peak Sunlight	5.3	
Total kW Capacity	100	
Daily Power Generation	530 kWh	
Annual	193450 kWh	
Electricity Rate	$0.18	

$0.18	0.185	0.189	0.194	0.199	0.204	0.209	0.214	0.219	0.225	0.230	0.236	0.242	0.248	0.254

Source: Prepared by Marty Silber, Consultant, Washington, DC

FIGURE 6-3 CONTINUED

Inputs	Watt
System Cost	5.20 Per watt
Rebates	2.60 Per watt
Net PV Cost	2.60
# of Watts	100000
Cost	$260,000
Peak Sunlight	5.3
Total kW Capacity	100
Daily Power Generation	530 kWh
Annual	193450 kWh
Efficiency	72 percent
Starting Electricity Rate	$0.18

Source: Prepared by Marty Silber, Consultant, Washington, DC

Figure 6-2 relates to a small 10 kW wind turbine installed in good wind conditions in a region with relatively high retail electric bills. Its revenues are derived from offsetting (or net metering) electricity starting at 18 cents per kWh and increasing 2.5 percent per year. No REC sales are assumed. Financing terms are 5 percent over 15 years.

Figure 6-3 relates to a 100 kW roof-mounted solar photovoltaic (PV) project installed in an area with high retail electricity costs. Its revenues are from offsetting (or net metering) 18 cents per kWh of retail electricity that increases in price 2.5 percent per year. Additional revenues come from a 50 percent state rebate and the sale of RECs at 75 cents per kWh. Financing terms are assumed to be 5 percent over 15 years. The analysis shows a net positive cash flow of $5,579. However, most colleges pay significantly less than 20 cents per kWh for their electricity even during peak demand periods. If a lower electricity price is used, the PV system's net cash flow becomes negative unless some other form of subsidy or cost reduction can be found.

▶ OWNERSHIP OPTIONS

Most colleges and universities begin on the assumption that they will own and operate their renewable energy projects. But there are alternative ownership approaches that can reduce costs, limit exposure to risk, provide price predictability, and allow for larger projects to be built than might otherwise be possible with sole ownership.

College or University Ownership

The simplest approach is for the college or university to own and operate the renewable energy project. The advantage is complete control of the project and first-hand knowledge of the costs, revenues, and technical performance. The disadvantage is the inability to make use of available tax benefits. Also, budgetary limitations may mean that the college or university will have to build a smaller project than it would have built with community partners or private investors.

Vendor Ownership

Under some ownership structures, the equipment provider retains ownership of renewable energy equipment, at least during an initial contract term. This is the case, for example, under leases, performance contracts, and the third-party service model. For a college or university, the obvious advantage of vendor ownership is not having to provide the capital for equipment purchase. Although the institution may eventually purchase the equipment, it does not have to raise a down payment. Under a third-party service arrangement, Stephens Institute of Technology in New Jersey had a 125 kW solar PV system installed at no cost. Stephens pays only for the solar electricity generated.

Vendor ownership is also advantageous when the energy equipment is innovative or experimental. In such cases, the vendor may want to retain ownership so that it can better monitor performance and make adjustments or repairs. The performance risk is mitigated for the college or university, and it faces no operations and maintenance costs. When a fuel cell was installed at SUNY College of Environmental Science and Forestry, for example, the vendor chose to retain ownership in order to engage in joint research with the university on the fuel cell's performance.

Private Equity Ownership

Private equity ownership projects, originally known as the Minnesota-style flip structure, are organized as for-profit limited liability companies (LLC) or corporations (see chapter 4). The college or university partners with private equity investors who are able to take advantage of depreciation and renewable energy tax benefits. Debt is provided either from a commercial bank or from the college itself; Carleton College, for example, provided a loan from its endowment. The college or university contributes as little as 1 percent of the equity for the project, and the private investors contribute 99 percent. During the first 10 years, the private investors receive 99 percent of the cash flow and 100 percent of the tax benefits. Revenues from electricity sales are used to pay debt service and profit to the equity investors. In year 11 (possibly later, if necessary), the equity investors sell the renewable energy project at fair market value to the college, and the ownership flips to 99 percent for the college and 1 percent for the investors. Now the college, having put little or nothing down, owns a debt-free wind project and receives 99 percent of the cash flow for the remaining life of the project.

Community Ownership

A community ownership structure involves establishing a partnership with other public and private institutions in the area to cooperatively purchase or invest in renewable energy. It is a vehicle for leveraging the college or university's investment and spreading the costs and risks so that a larger project can be built. It also can foster goodwill between the institution and the broader community.

Saint Francis University formed a renewable energy consortium with local businesses and nonprofit organizations in western Pennsylvania that are interested in buying clean energy. The consortium includes schools, hospitals, and a prison. So far, the consortium is acting as a buyers' cooperative to find bulk purchase deals on energy-efficiency products and services and wind power RECs. Now it is considering participating in the development and ownership of a wind farm.

Some community partners are more advantageous than others. Having the local municipal or county government as a member can give the partnership access to bond financing and can expedite approvals and permitting. Successful businesses with an interest in tax advantages will allow the partnership to benefit from renewable energy tax incentives. To maximize the tax advantages, these businesses may need to form their own corporate entity, which would in turn join the community partnership. It may be necessary for tax purposes for the private entity to own the renewable energy facility until the tax benefits expire, after which the private entity would flip the ownership over to the community owners.

A college or university can partner with farmers or other landowners who have suitable wind conditions for wind turbine sites. A relatively new concept is codigestion, or partnering with farmers for a supply of farm and animal waste for a community anaerobic digester for methane biogas production. Such a partnership will increase biogas production and reduce tipping fees and hauling fees for waste disposal.

The following industries commonly have the need to dispose of high-strength organics, either onsite or by offsite disposal or land application systems:[2]

- Dairy processing, including ice cream, cheese, yogurt, sour cream, and milk condensing and bottling
- Brewing and beverages, including beer, distilled spirits, wine, juice, and soda
- Nutraceutical production (fortified foods and supplements)
- Fruit and vegetable processing
- Prepared foods production, from frozen meals to salad dressing

The down side of community ownership is procedural bureaucracy. Each member has its own budgetary, procurement, and investment guidelines, as well as timetable and approval processes, which are difficult to coordinate.

▶ CONNECTING TO THE GRID

All projects that generate electricity for campus facilities—even those that will not be exporting electricity to the local electric utility—require approvals from the utility to avoid power quality problems. The utility does not need to approve generation projects that are providing power to an isolated facility, such as a building on a research farm that is unconnected to the campus electrical grid.

A college or university may decide it wants to export some portion of its generated power to the utility and engage in power sales. It will thus have to negotiate the price and terms under which the utility will buy the renewable energy offered to it. As part of this negotiation, the institution may want (or need) to negotiate possible changes in the price and terms under which it continues to buy electricity from the utility.

Many of the issues involved in connecting or "interconnecting" to the utility are addressed in state law. Most states have statutes or regulations that deal with utility purchases of renewable energy. These laws have been needed because utilities typically prefer not to buy power from small power providers. Even when required, utilities can and do place what can be costly conditions on the purchase. They can also end up offering far less than the price they pay for power from conventional power plants and certainly lower than the price at which the college or university is buying its electricity. The college or university should be prepared for some resistance and negotiation.

Even when the college or university is using all of its renewable power on site, it may encounter difficulties. When Pierce College installed a solar photovoltaic system in conjunction with a new CHP system, the Los Angeles Department of Water and Power added an annual standby charge of $50,000 to the college's electric bill (see chapter 3 on standby charges). The utility claimed the charge was justified because the college's clean energy units might break down and the utility would need to have standby generation.[3] This example is a case of one government agency—a municipal utility—creating a financial barrier to renewable energy even as another agency—the state public utility commission—was offering financial incentives to the college to invest in renewable energy.

Some owners of self-generation projects have been told at the outset that there would be no standby charge only to find that the utility decides to assess one in later years. Even when there are no standby charges, the institution may want to include a clause in its utility interconnection agreement that there never will be a standby charge. States are beginning to regulate standby charges, and some disallow them for renewable energy projects.

Power Purchase Agreement

A college or university will enter into two agreements with the utility: a power purchase agreement and an interconnection agreement. The power purchase agreement (PPA) is only necessary if the institution is selling some or all of its renewable electricity to the utility. The PPA specifies how much electricity the utility will buy, over what time period, and either the price or the method of calculating that price. Even if utilities are required by law to purchase the college or university's electricity, there is still a lot to be negotiated. Utilities generally oppose PPAs with small power producers and resist paying a good price for the electricity.

In cases where a college or university plans to rely on the utility's payments to help service the debt on a renewable energy project, they should be aware that banks will generally not finance such a project without a 10-year or longer PPA. This is applicable mainly to wind projects, but it may apply to other projects as well. The utility will often press for a much shorter PPA period.

In states that require net metering, there is no need to negotiate the electricity price because the utility is required to buy the college or university's electricity at the rate at which the institution is buying electricity (see the discussion of net metering in chapter 2).

In some cases, PPAs address the disposition of RECs. Colleges and universities should make sure they are not inadvertently transferring ownership of the RECs that accrue to them from their renewable energy project.

Interconnection Agreement

An interconnection agreement is needed for all college and university renewable energy projects, even those that do not export electricity off campus. This technical agreement commits the institution to generating electricity in a manner that does not adversely affect the safe and reliable operation of the utility's electricity distribution system. The utility will want the college or university to avoid the generation of harmonic frequencies, voltage fluctuations, or operating at voltage levels outside normal ranges. These phenomena can cause problems with the utility's equipment and can damage the lights, appliances, and other equipment of its other customers. In the interconnection agreement, the utility will often provide specifications (and even the brand name and model number) of the equipment needed and will require that the college or university conduct specific testing and maintenance activities in order to avoid the creation of power quality problems in the future.[4]

In some cases, the utility may express concerns that merely specifying equipment and maintenance practices is insufficient to protect its system. The utility may thus decide to conduct a thorough technical appraisal of the college or university's proposed plant and associated equipment. The appraisal is known as a power quality impact analysis or interconnection study and can cost $50,000 or more. The institution must bear this cost.

Colleges and universities should avoid an interconnection study, not just because of the high cost, but because the study could lead the utility to impose operational requirements that add procedures and costs to the projects. Operational requirements are included in the interconnection agreement and specify when the plant can and cannot supply electricity to the utility grid.

For small systems, interconnection studies are unnecessary as long as the college or university agrees to install the specified equipment. The definition of what is small in this situation is a project that is less than 15 percent of the utility's circuit. The utility's circuit refers to the particular transformer in the local substation that serves the college or university. The capacity of transformers generally ranges from 10 MW to 40 MW. Thus, for a project to reach 15 percent of the smallest circuit (a 10 MW circuit), it would have to be at least .15 x 10, or 1.5 MW in size. To date, the majority of college and university renewable energy projects are significantly smaller.

However, some utilities claim that even a project smaller than 15 percent of the circuit—and as small as 1 percent—can cause problems and will necessitate an interconnection study and the associated expense. Colleges and universities encountering this situation should be aware that the 15 percent guideline has been adopted as a federal rule applicable to generators interconnecting to federal government utilities. It has also been adopted as a rule at the state level in California, New Jersey, and Ohio, while Arizona, Colorado, and Pennsylvania are in the process of adopting it. The 15 percent guideline is a widely accepted standard, including by the utilities' trade association, the Edison Electric Institute.

In essence, colleges and universities should be aware that some utilities seek to derail renewable energy projects by making them more costly. One way of adding to the costs is to require the institution to pay for an interconnection study based on the claim that the project could cause power quality problems for the utility. Colleges and universities must thus be cognizant of the 15 percent rule and be able to show that the proposed project is too small to warrant an interconnection study.

FIGURE 6-4: THREE APPROACHES TO UTILITY INTERCONNECTION

On-Site Consumption

Institution consumes all renewable power on campus.

Institution pays utility for remaining power at normal retail rate, possibly faces standby charges.

Wholesale Sale

Utility pays institution for renewable power at a *wholesale* rate.

Institution pays utility for power at normal retail rate, faces no standby charges.

Retail Sale (Net Metering)

Utility pays institution for renewable power at *retail* rate.

Institution pays utility for power at normal retail rate, faces no standby charges.

Source: Michael Philips

▶ PRECONSTRUCTION ARRANGEMENTS

Once the resource analysis and site assessments, initial economic analysis, and determination of ownership structure have been completed, and as the discussions begin with the utility regarding interconnection and electricity sales, the next step is project design and preconstruction. Every project faces its own special set of issues, but the following general considerations will help the college or university prepare to build the project. Many steps will be reduced or even eliminated if a performance contract or third-party service approach is chosen (see chapter 5).

Design Work

In most cases, design work is handled by an engineering, procurement, or construction (EPC) contractor in consultation with facilities staff. For simple projects such as small or medium-sized rooftop solar PV projects, some colleges and universities do the design work in-house, using some outside consulting for decisions on specific components or system integration. In other cases, the equipment vendor handles the design work. California State University Northridge, for example, included the design work in its RFP to equipment vendors. The university processed a $900,000 solar PV equipment order (for a $2 million PV project) and paid the vendor an extra $40,000 for design work, or an additional 2 percent on top of the equipment and installation cost. Still other colleges and universities combine their PV installations with other campus energy work, such as energy management or installation of conventional power generation, and hire consulting engineers for design. In some cases, engineering students actively participate in the design work.

In general, large wind projects involve the most extensive and expensive design. They require site planning, underground wiring configurations, and soil borings to determine foundation design. Solar projects are generally the easiest and least costly in terms of design. It is difficult to generalize about biomass, but typically biomass projects that do not involve a fuel conversion step (e.g., converting solid biomass to gas) entail a more straightforward design phase. The easiest and quickest designs are for biomass cofiring, landfill methane recovery, and direct combustion of woodchips for space heating.

Studies

Projects may require several types of preliminary studies. Before signing an interconnection agreement, an engineering consultant will conduct an interconnection analysis or power quality impact analysis. This analysis, which is much smaller than the interconnection study described earlier in this chapter, examines the design features of the renewable energy project and its interconnection, models the project's voltage profile and power quality impacts, and suggests any necessary design modifications to avoid negative impacts. A permitting appraisal will determine what kinds of permits will be needed and their costs. The permits typically will involve some combination of electrical, building, and special use permits. In some cases, permitting is standardized. Additional studies include a telecommunications impact assessment (for wind towers), an environmental impact assessment, soil analysis to determine the suitability for ground-penetrating structures such as wind towers or PV stanchions, and a historic preservation, cultural resources, or archeological survey (mainly for off-campus projects).

Warranties and Guarantees

The standard warranty on solar PV panels is 20 years. The standard for inverters is 5 years, although inverters typically last 10 years or more. In addition to a product warranty, the college or university may wish to have a performance guarantee. It is difficult to obtain performance guarantees for wind turbines because it is impossible to predict future wind strengths. When obtaining a PV guarantee, the college or university may request that the supplier of the PV modules or the third-party service provider guarantee a minimum output performance from the solar system over the course of a calendar year. The California Power Authority provided such a performance guarantee in a 2004 RFP for providers of solar PV systems to colleges, universities, and other state institutions.

The RFP required each bidder to state a "quantity of power they expect to deliver each year." Each bidder was also asked to guarantee "a minimum output performance from the solar system over the course of each calendar year, at a minimum level equal to 90 percent of the stated expected performance output."[5]

Some states have a certification procedure for solar and other renewable energy equipment. Buyers are not required to procure only certified equipment, but eligibility for state financial incentives may depend on purchasing only state-certified equipment. If a contractor will handle equipment procurement, colleges and universities may want to make sure the procured equipment is state-certified.

ENDNOTES

1.   See www.nrel.gov/gis/solar_maps.html.

2.   John Ewing, "Agricultural Anaerobic Digestion: General Overview," presentation at Anaerobic Digester & Methane Recovery Workshop, October 5, 2004, Morgan County REA, Fort Morgan, CO, available at: www.state.co.us/oemc/events/anaerobic/index.html.

3.   Blair Doane, regional facilities project coordinator, Los Angeles Community College District, personal communication, June 2005.

4.   The equipment must meet standard IEEE 1547. Any UL-certified equipment will by definition be in compliance. For example, the inverters purchased for a solar photovoltaic project should be UL-certified.

5.   State Consumer Power and Financing Authority, RFP of April 22, 2004, available at: www.capower-authority.ca.gov, p. 18.

CONCLUSIONS AND
NEW DIRECTIONS

As environmental, economic, and financial trends converge, colleges and universities are being encouraged to meet a significant portion of their energy demand with renewable sources of energy. Students and faculty across the United States are urging their institutions to buy renewable energy as a way to address global warming. Those institutions are facing rising energy prices and new concerns about energy security and self-reliance. The public is concerned about the economic and environmental consequences of the nation's over-reliance on imported fossil fuels and is looking for leadership to reduce that reliance.

Over the last 20 years, many colleges and universities have installed small solar energy demonstrations. But in recent years, a number of them have gotten serious about renewable energy and are either installing large solar, wind, and biomass projects, or are purchasing 10 percent or more of their electricity from renewable energy suppliers.

Although the price per kilowatt-hour of most renewable energy sources is higher than fossil fuel prices, renewable energy investments can be economical under many circumstances. The availability of government incentives, hedge products, and creative financing approaches and ownership schemes means that renewable energy can be affordable and even profitable for many colleges and universities.

▶ RECS

Purchasing renewable energy credits (RECs) from the local electric utility or a renewable energy supplier is the most immediate step a college or university can take to support clean energy. Many colleges and universities buy RECs equivalent to 5–10 percent or more of their electricity consumption. At a price of 2 cents per kWh or less, RECs are the lowest-cost form of renewable energy. The premium the institution pays supports the construction of new renewable energy projects, and the emissions reductions a REC represents can be applied to a campus greenhouse gas reduction strategy and recognized as such by the U.S. Environmental Protection Agency's Green Power Partnership Program. For colleges and universities that are pursuing green building certification by the U.S. Green Building Council, a REC purchase can earn low-cost LEED points.

▶ HEDGING FOSSIL FUEL PRICE VOLATILITY

Although buying RECs has advantages, it also results in higher energy prices because the college or university typically pays for the certificates on top of the electricity it already purchases from the power grid. A variation on a straight REC purchase is the purchase of renewable energy on a long-term fixed-price basis, thereby hedging the price volatility of fossil fuel-based grid power. As in a straight REC purchase, the college or university pays a premium for the renewable energy, but it acquires a hedge against future grid electricity price increases. If grid prices increase over time, the hedge purchase yields considerable savings, as in the case of Concordia University in Austin, Texas. If the utility does not offer long-term fixed-price contracts for renewable energy (and few do), the college or university can opt for a contract for differences with a renewable energy supplier, which creates a hedge against grid electricity price increases (see chapter 5 for a case study of Concordia University and a discussion of contracts for differences).

▶ COMBINING REC PURCHASES WITH ON-CAMPUS PROJECTS

The disadvantage of buying renewable energy through suppliers—either as RECs or as hedge products—is the absence of a physical, on-campus project to show for it. It is a purely financial transaction. Ideally, a college or university may decide that the best course of action is a combination of RECs and on-site power generation. Alternatively, it may decide to start with a REC purchase and graduate to an on-site project.

In general, the on-site renewable energy sources with the lowest cost per kWh basis are landfill methane combustion and solid biomass combustion, assuming that the campus is near a municipal landfill or a supply of biomass feedstock. Even less costly—for colleges and universities that have conventional on-site fossil fuel power generation—is cofiring fossil fuel with some form of biomass, such as wood or agricultural residues.

Large-scale wind power is economically viable in some areas of the country, and some colleges and universities in Minnesota and Iowa are earning good returns from their wind investments. If sufficient federal and state financial incentives are available, a solar photovoltaic (PV) project can be made economically viable.

Solar PV installations are currently the most common form of renewable energy on campuses. The sunlight resource is available almost everywhere, installation space is available on rooftops and parking lots, and systems can be installed on a small scale and expanded on a modular basis as additional capital resources become available. Although solar PV costs more per kilowatt-hour than most other renewable energy sources, some innovative financing approaches are available.

From a purely financial viewpoint, the best returns for a college or university per dollar invested will come from equity investments in wind farms. The wind farms may be thousands of miles from campus, and the institution may not actually receive power from them. Equity investors in wind farms used to be an exclusive club, but this situation has changed, and now there are opportunities to coinvest with private senior partners. The minimum investment may be sizeable, so a college or university may want to form a small investor consortium, possibly with other interested institutions.

▶ FINANCING RENEWABLE ENERGY PROJECTS

All but the smallest on-campus renewable energy projects usually involve some form of debt. State university systems and community college districts that have included them as components of bond issues report that voters are highly supportive of renewable energy projects. At least one institution, Carleton College, has provided a loan from its endowment for its renewable energy project. Leasing is less common, although lease-like performance contracts and third-party supplier models are on the rise. Their attractiveness lies in the lack of required down payment, predictability of cash flow, and avoidance of technology performance risk. Several California colleges and universities have entered into third-party supplier contracts for the installation of solar PV systems. Federal and state grants and rebates are available for most forms of renewable energy and are an essential financial component of nearly every project.

Some colleges and universities earn returns from selling their electricity to the local electric utility. This option is particularly viable in states that allow net metering. It also may be feasible in states with renewable portfolio standard laws, which require the utilities to derive a certain percentage of their electricity supply from in-state renewable energy projects. Some universities—such as California State University, Hayward—sell the RECs that accrue to them from renewable energy projects to help pay for the projects.

▶ UTILITY INTERCONNECTIONS

All on-campus electricity-generating projects must have interconnection agreements with the local distribution utility, even if the projects do not export electricity from the campus. For this reason, utilities become involved in all projects, at least peripherally. When the college or university exports power, the utility becomes involved more directly. Some utilities seek to assess standby charges (see chapter 3) or, usually for larger projects, require expensive interconnection engineering assessments. Such charges can sink the financial viability of a renewable energy project. It is best to work with an expert who can fend off or at least minimize utility attempts to add costs to a project.

▶ INTEGRATING RENEWABLE ENERGY WITH OTHER INITIATIVES

Renewable energy investment should be incorporated as a component of the college or university's long-term capital plans and budgets, energy plans, environmental plans, and security plans.

Capital Planning and Budgeting

Renewable energy projects should be integrated with the college or university's strategic planning process, which in turn is often linked to its operating and capital budgeting processes. Instead of regarding a renewable energy initiative as a one-time project, its inclusion in the strategic and capital planning processes will put the college or university in a stronger position to consider financing opportunities and options and build them into its capital development planning. Ideally, renewable energy projects will be integrated with plans for campus growth, renovation, and modernization. As a result, the broader college community will take ownership of the project instead of regarding it as the pet project of the facilities, engineering, or environmental studies department. When renewable energy projects are incorporated into planning, their financing can be bundled with financing for broader capital investments. This approach also ensures that new construction will incorporate renewable energy design features.

Energy Planning

The best approach may be to invest in a combination of grid-supplied renewable energy kWh (or RECs) and on-site renewable energy generation. The on-site installation gives the institution visibility, security, and a level of self-reliance, while grid-supplied renewables give it more bang for the buck—more kWh of renewables per dollar invested—and this is a greater environmental benefit. Both scenarios offer a hedge against future price volatility of conventional fuels, if done correctly. Harvard University is doing both: Currently Harvard's green power purchases and on-site generation add up to over 21,692 MWh, which accounts for more than 7 percent of the total electrical load.[1]

Environmental Planning

In response to increasing attention to carbon dioxide and global climate impacts, universities like Tufts are adopting climate policies. But the more localized pollutants like nitrogen oxides (NOx), particulate matter (PM), and mercury should not be overlooked. As state governments seek to comply with stricter EPA rules on NOx and PM, colleges and universities can play a role in reducing these emissions through their energy-efficiency and renewable energy initiatives and can get credit and even funding from state environmental agencies. Note that getting credit and funding will require submitting more than simple estimates of emissions reductions. The EPA will require the use of a sophisticated model like eCalc to accurately calculate the emissions reductions resulting from individual energy-efficiency and renewable energy measures. The models are not expensive, but they do take some time to learn and use, and training courses are available.

Unlike NOx reductions, carbon emission reductions currently have little or no monetary value in the U.S. There have been some cross-border sales with Canada, but the price paid is very low because there is a far greater supply of carbon reductions or credits than there are buyers. Carbon trading is at an early stage in Europe and it will be interesting to see if colleges and universities get into the act.

Sustainability initiatives—Oberlin College (OH) and Berea College (KY) have renewable energy systems as part of a larger set of sustainability initiatives often implemented in a specific building or set of buildings.

Security Planning

A college should ensure that its on-site renewable energy system, if not providing 100 percent of the campus's electricity, is wired so that during blackouts it supplies power to the campus's critical functions such as emergency lighting, the medical clinic, campus security and fire facilities, critical computer and telecommunications systems, and minimum water and sewage pumping.

▶ NEW DIRECTIONS

Initiate Community Projects

Co-financing and co-owning projects with other colleges, K–12 schools, municipalities, and nonprofit organizations is an underutilized practice. Farmers are starting to cooperatively co-finance wind turbines and dairy methane recovery projects. Carlton College's wind project started out as a community effort. There will inevitably be problems coordinating all the institutions and their approvals and rules for expenditures, but the advantage is that larger projects can be undertaken. For biomass projects, it may be possible to enter into partnerships with farmers who can supply the biomass product (e.g., crop waste, dairy waste, etc.). For wind projects where the best wind conditions are off-campus on agricultural land, farmers can be partners, not just land lessors. An added benefit is that they have access to USDA funds for renewable energy.

Bundle On-Campus Projects with Other Colleges & Universities

Under the umbrella of the New Jersey Higher Education Partnership for Sustainability (NJHEPS), New Jersey colleges and universities have coordinated studies of their carbon emissions and plans for reducing them. A consortium of colleges could take this a step further by forming a buyer's cooperative for energy-efficiency and renewable energy equipment. In a region with good wind conditions, such as the Midwest or West, the cost of wind projects could be lowered by coordinating the timing of projects and standardizing technical specifications and bidding documents. Each project would be unique in terms of its PPA and utility interconnection agreements, but there could be a better EPC price from engineering and construction contractors if they are bidding on a package of projects instead of just one. There may also be a way to bundle the projects for financing, although there may not be any advantage to doing that for projects seeking private investment because neither equity investors nor commercial banks will buy into a blind trust. They will want to examine each project separately, so there will be little or no savings, particularly on transaction costs. However, there may be some merit in bundling smaller wind projects (>100 KW) for the Department of Agriculture's 25 percent grant application for state programs.

Invest in Renewable Energy Projects and Enterprises Off-Campus

At one time, there were barriers to colleges and universities taking equity stakes in wind farms. That began to change in 2004 and now it is possible for these institutions to invest directly in large-scale wind farms. Endowments or capital budget funds could be used to purchase shares of a wind farm. Advantages include financial returns from electricity sales, and access to low-cost RECs, which can be owned as part of the college's greenhouse gas reduction strategy or sold for additional financial returns.

A group of colleges could jointly provide the equity for a wind farm. The colleges do not need to be in the same vicinity, but could simply share an interest in buying renewable energy at the best possible price, with no middleman institutions like brokers, utilities, REC suppliers, etc.

The colleges can direct their local utility to wheel the electricity from the wind farm to their campus and include it as part of their electricity resource package. This will work particularly well in states where there is retail competition among electric power suppliers because a college can specify what power plants it wants to buy its power from and then ask for proposals from the various power suppliers on wheeling costs and other costs.

Establish a REC Clearinghouse/Exchange

This option allows colleges and universities to sell, buy, and trade RECs. RECs would be less extensive and the money for them would go to another institution instead of to a utility or REC broker. This could be broadened to encompass all nonprofit institutions buying and selling RECs, such as municipal governments, K-12 schools, and hospitals. This could be coordinated by some of the national or regional college collaboratives such as NJHEPS.

ENDNOTES

1.   EPA Green Power Partnership, www.epa.gov/greenpower/partners/partners/harvarduniversity.htm.

SELECTED COLLEGES AND UNIVERSITIES WITH RENEWABLE ENERGY INVESTMENTS

1. Purchases of green power from utilities or REC suppliers

California

California State University System
Santa Clara University
Stanford University
University of California System

Colorado

Colorado State University
Naropa University
University of Colorado, Boulder

Connecticut

Connecticut College
Wesleyan University
Yale University School of Forestry and Environmental Studies

District of Columbia

American University
Catholic University of America

Idaho

University of Idaho

Maine

Colby College
College of the Atlantic
Unity College
University of Maine, Orono
University of Southern Maine

Massachusetts

Harvard University
Tufts University

Michigan

University of Michigan, Ann Arbor

New Jersey

Kean University
Monmouth University
Montclair State University
Richard Stockton College of New Jersey
Rutgers University
William Patterson University

New York

Bank Street College of Education
City University of New York
Hamilton College
Hobart and William Smith Colleges
Pace University
Rockland County Community College
Sienna College
Syracuse University
Union College
University at Buffalo, State University of New York

North Carolina

Appalachian State University
Duke University
University of North Carolina, Chapel Hill

Ohio

Oberlin College

Oklahoma

University of Oklahoma

Oregon

Lewis and Clark College
Oregon State University
University of Portland

Pennsylvania

Allegheny College
Bloomsburg University
Bucknell University
California University of Pennsylvania
Carnegie Mellon University
Chatham College
Cheyney University
Clarion University
Dickinson College
Drexel University
Duquesne University
East Stroudsburg University
Eastern University
Edinboro University
Franklin and Marshall College
Gannon University
Gettysburg College
Haverford College
Indiana University of Pennsylvania
Juniata College
Keystone College
Kutztown University
Lock Haven University
Mansfield University
Mercyhurst College
Messiah College
Millersville University
Pennsylvania State University
Shippensburg University
Slippery Rock University
Swarthmore College
Temple University
University of Pennsylvania
West Chester University

South Carolina

Coastal Carolina University

Tennessee

Maryville College
University of Tennessee, Knoxville

Texas

Concordia University
Texas Wesleyan University United Methodist Campus Ministry

Utah

University of Utah

Washington

Bainbridge Graduate Institute
Evergreen State College
Western Washington University
Whitman College

Wisconsin

University of Wisconsin, Oshkosh

2. On-Site Renewable Energy Systems (of at least 12 KW in size)

California

California State University Hayward	1 MW PV
California State University Northridge	462 kW PV and 225 kW PV
Cerro Coso Community College	1 MW PV
Glendale Community College	400 kW PV
Pierce College	191 kW PV
UCLA	Landfill methane-powered generator

District of Columbia

Georgetown University	300 kW PV

Illinois

Art Institute of Chicago	44 kW PV

Iowa

Iowa Lakes Community College	1.65 MW wind
University of Iowa	Oat hull (biomass) cofiring with coal

Maine

Colby College	Campus is 100 percent fueled by hydro and biomass

Maryland

Montgomery College, Germantown	25.6 kW PV
Montgomery College, Rockville	26 kW PV

Massachusetts

Harvard Business School	36.5 kW PV
Mount Wachusett Community College	Wood biomass heating system
Northeastern University	18 kW PV

Michigan

Aquinas College	26.1 kW PV
Calvin College	20 kW PV
Central Michigan University	Woodchip heating and cooling plant

Minnesota

Carlton College   1.65 MW wind

Mississippi

University of Mississippi   29.4 kW PV on intramural sports
                                                         complex

Nebraska

Chadron State College   Biomass (forest residues) for
                                                         campus heating plant

Nevada

University of Nevada, Las Vegas   28 kW PV

New Jersey

Kean University   28 kW PV
Monmouth University   454 kW PV
New Jersey Institute of Technology   50 kW PV
Ramapo College   100 kW PV
Stevens Institute of Technology   125 kW PV
Stockton College   18 kW PV

New York

Hudson Valley Community College   Landfill methane-powered
                                                         generator

SUNY, College of Environmental
  Science and Forestry   17 kW PV
SUNY, Morrisville State College,   50 kW dairy manure–fueled
                                                         gasifier-generator; 10 kW wind

North Carolina

Catawba College   Solar PV and solar hot water

North Dakota

Grafton Technical College   70 kW wind

Ohio

Oberlin College   60 kW PV & geothermal
                                                         integrated with other sustainability
                                                         features in one building

Oregon

University of Oregon   12 kW PV

Texas

University of Texas, Austin	24 kW PV
University of Texas, Health Science Center	45.3 kW PV
University of Texas, Medical Branch	20 kW PV

Virginia

University of Virginia	50 kW PV

Wisconsin

Nicolet College	10 kW wind; two PV installations
Northland College	Combined solar-wind installation

Wyoming

University of Wyoming	35 kW PV

Canada

Trent University, Canada	Small hydro

RENEWABLE ENERGY CERTIFICATE RETAIL PRODUCTS

(as of October 2005)					
Certificate Marketer	**Product Name**	**Renewable Resources**	**Location of Renewable Resources**	**Residential Price Premiums***	**Certification**
3 Phases Energy Services	Green Certificates	100% new wind	Nationwide	2.0¢/kWh	Green-e
Blue Sky Energy Corp	Greener Choice™ Green Tags	Landfill Gas	Utah	1.95¢/kWh	—
Bonneville Environmental Foundation	Green Tags	≥98% new wind, ≤1% new solar, ≤1% new biomass	Washington, Oregon, Wyoming, Montana, Alberta	2.0¢/kWh	Green-e
Clean Energy Partnership/ Community Energy	Mid-Atlantic Wind	100% new wind	Mid-Atlantic	2.0¢/kWh	Green-e
Clean Energy Partnership/ Sterling Planet	National New Clean Energy Mix	24% wind, 25% bio-mass, 50% landfill gas, 1% solar	Nationwide	0.6¢/kWh	Environmental Resources Trust
Clean Energy Partnership/ Sterling Planet	National and Regional New Wind	100% new wind	Nationwide	1.0¢/kWh	Environmental Resources Trust
Clean and Green	Clean and Green Membership	100% new wind	Nationwide	3.0¢/kWh	Green-e
Community Energy	New Wind Energy	100% new wind	Colorado, Illinois, New York, Pennsylvania, West Virginia	2.0¢/kWh - 2.5¢/kWh	Green-e
Conservation Services Group	ClimateSAVE	95% new wind, 5% new solar	Kansas (wind), New York (solar)	1.65¢/kWh - 1.75¢/kWh	Green-e
EAD Environmental	100% Wind Energy Certificates	100% new wind	Not specified	1.5¢/kWh	—
EAD Environmental	Home Grown Hydro Certificates	100% small hydro (<5MW)	New England	1.2¢/kWh	—
Green Mountain Energy	TBD (Pennsylvanian REC product)	100% wind	Nationwide	1.7¢/kWh- 2.0¢/kWh	—
Maine Interfaith Power & Light/BEF	Green Tags (supplied by BEF)	≥98% new wind, ≤1% new solar, ≤1% new biomass	Washington, Oregon, Wyoming, Montana, Alberta	2.0¢/kWh	—
Mass Energy Consumers Alliance	New England Wind	100% new wind	Massachusetts	5.0¢/kWh	—
NativeEnergy	CoolHome	New biogas and new wind	Vermont and Pennsylvania (biomass), South Dakota (wind)	0.8¢/kWh - 1.0¢/kWh	**

Certificate Marketer	Product Name	Renewable Resources	Location of Renewable Resources	Residential Price Premiums*	Certification
NativeEnergy	WindBuilders	100% new wind	South Dakota	~1.2¢/kWh, $12 per ton of CO_2 avoided	**
Renewable Choice Energy	American Wind	100% new wind	Nationwide	2.0¢/kWh	Green-e
Renewable Ventures	PVUSA Solar Green Certificates	100% solar	California	3.3¢/kWh	Green-e
SKY energy, Inc.	Wind-e Renewable Energy	100% new wind	Nationwide	2.4¢/kWh	Green-e
Sterling Planet	Green America	45% new wind, 50% new biomass, 5% new solar	Nationwide	1.6¢/kWh	Green-e
TerraPass Inc.	TerraPass	Various (including efficiency and CO_2 offsets)	Nationwide	~$11/ton CO_2	—
Waverly Light & Power	Iowa Energy Tags	100% wind	Iowa	2.0¢/kWh	—
WindCurrent	Chesapeake Windcurrent	100% new wind	Mid-Atlantic States	2.5¢/kWh	Green-e

Notes

* Product prices are updated as of June 2005. Premium may also apply to small commercial customers. Large users may be able to negotiate price discounts.

** The Climate Neutral Network certifies the methodology used to calculate the CO_2 emissions offset.

NA = Not applicable.

Source: This table has been reprinted from National Renewable Energy Laboratory Technical Report NREL/TP-38994, Green Power Marketing in the United States: A Status Report, 8th ed. *(October 2005) by Lori Bird and Blair Swezey.*

RETAIL GREEN POWER PRODUCT OFFERINGS

(as of October 2005)

State	Company	Product Name	Residential Price Premium[1]	Fee	Resource Mix[2]	Certification
CT	Community Energy (CT Clean Energy Options Program)	CT Clean Energy Options 50% or 100% of usage	1.1¢/kWh	—	50% new wind, 50% landfill gas	—
CT	Levco	100% Renewable Electricity Program	0.0¢/kWh	—	98% waste-to-energy and hydro (Class II), 2% new solar, wind, fuel cells, and landfill gas	—
CT	Sterling Planet (CT Clean Energy Options Program)	Sterling Select 50% or 100% of usage	1.15¢/kWh	—	33% new wind, 33% existing small low-impact hydro, 34% new landfill gas	—
DC	PEPCO Energy Services (3)	Green Electricity 10%, 51% or 100% of usage	1.35¢/kWh (for 100% usage)	—	landfill gas	—
DC	PEPCO Energy Services (3)	NewWind Energy 51% or 100% of usage	2.05¢kWh (for 100% usage)	—	new wind	—
DC	Washington Gas Energy Services/Community Energy	New Wind Energy (5%, 10%, 25%, 50%, or 100% of usage)	2.5¢/kWh	—	new wind	—
ME	Maine Renewable Energy/Maine Interfaith Power & Light (4)	Maine Clean Power	2.37¢/kWh	—	100% low-impact hydro	—
ME	Maine Renewable Energy/Maine Interfaith Power & Light (4)	Maine Clean Power Plus	2.87¢/kWh	—	80% low-impact hydro, 20% wind	—
MD	PEPCO Energy Services (5)	Green Electricity 10%, 51% or 100% of usage	2.75¢/kWh (for 100% usage)	—	landfill gas	—

State	Company	Product Name	Residential Price Premium[1]	Fee	Resource Mix[2]	Certification
MD	PEPCO Energy Services (5)	NewWind Energy 51% or 100% of usage	3.35¢/kWh (for 100% usage)	—	new wind	—
MD	PEPCO Energy Services (5)	Nonresidential product	NA	—	50% to 100% eligible renewables	Green-e
MD	Washington Gas Energy Services/Community Energy	New Wind Energy	2.5¢/kWh	—	new wind (5%, 10%, 25%, 50%, or 100% of usage) or 100 kWh blocks	—
MA	Cape Light Compact (6)	Cape Light Compact Green 50% or 100%	1.768¢/kWh (for 100% usage)	—	75% small hydro, 24% new wind or landfill gas, 1% new solar	—
MA	Massachusetts Electric/Nantucket Electric/Community Energy	New Wind Energy 50% or 100% of usage	2.4¢/kWh	—	50% small hydro, 50% new wind	Green-e
MA	Massachusetts Electric/Nantucket Electric/Mass Energy Consumers Alliance	New England Green-Start 50% or 100% of usage	2.4¢/kWh (for 100% usage)	—	75% small hydro, 19% biomass, 5% wind, 1% solar (≥25% of total is new)	—
MA	Massachusetts Electric/Nantucket Electric/Sterling Planet	Sterling Premium 50% or 100% of usage	1.35¢/kWh	—	50% small hydro, 30% bioenergy, 15% wind, 5% new solar	Environmental Resources Trust
NJ	PSE&G/JCP&L/Community Energy	Clean Power Choice Program	1.3¢/kWh	—	50% wind, 49% low-impact hydro, 1% solar	—
NJ	PSE&G/JCP&L/Green Mountain Energy	Clean Power Choice Program	0.9¢/kWh	—	50% wind, 50% low-impact hydro	—
NJ	PSE&G/JCP&L/Jersey-Atlantic Wind	Clean Power Choice Program: New Jersey Wind Energy	5.5¢/kWh	—	100-kWh new wind	—
NJ	PSE&G/JCP&L/Jersey-Atlantic Wind	Clean Power Choice Program	2.9¢/kWh	—	50% wind, 50% low-impact hydro	—
NJ	PSE&G/JCP&L/Sterling Planet	Clean Power Choice Program	1.2¢/kWh	—	33% wind, 33% small hydro, 34% bioenergy	Environmental Resources Trust
NY	Con Edison/Sterling Planet	NY Clean Choice	1.25¢/kWh	—	40% new wind, 30% small hydro, 30% landfill gas	Environmental Resources Trust
NY	ConEdison Solutions (8)/ Community Energy	GREEN Power	0.5¢/kWh	—	25% new wind, 75% small hydro	Green-e

State	Company	Product Name	Residential Price Premium[1]	Fee	Resource Mix[2]	Certification
NY	ECONnergy	Keep It Clean	$.10/day for 100kWh $.20/day for 200kWh	—	100% new wind	—
NY	Energy Cooperative of New York (9)	Renewable Electricity	0.5¢/kWh to 0.75¢/kWh	—	25% new wind, 75% existing landfill gas	—
NY	Long Island Power Authority/ Community Energy	New Wind Energy	2.5¢/kWh	—	new wind	—
NY	Long Island Power Authority/ Community Energy	New Wind Energy and Water	1.3¢/kWh	—	60% new wind, 40% small hydro	—
NY	Long Island Power Authority/ EnviroGen	Green Power Program	1.0¢/kWh	—	75% landfill gas, 25% small hydro	—
NY	Long Island Power Authority/ Sterling Planet	New York Clean	1.0¢/kWh	—	55% small hydro, 35% bioenergy, 10% wind	—
NY	Long Island Power Authority/ Sterling Planet	Sterling Green	1.5¢/kWh	—	40% wind, 30% small hydro, 30% bioenergy	—
NY	NYSEG/Community Energy	Catch the Wind/ New Wind Energy	2.5¢/kWh	—	100-kWh blocks of new wind	—
NY	Niagara Mohawk/Community Energy	60% New Wind Energy and 40% Small Hydro	1.0¢/kWh	—	60% new wind, 40% hydro	—
NY	Niagara Mohawk/Community Energy	NewWind Energy	2.0¢/kWh	—	new wind	—
NY	Niagara Mohawk/EnviroGen	Think Greenw	1.0¢/kWh	—	75% landfil! gas, 25% hydro	—
NY	Niagara Mohawk/Sterling Planet	Sterling Green	1.5¢/kWh	—	40% wind, 30% small hydro, 30% bioenergy	Environmental Resources Trust
NY	Niagara Mohawk/Green Mountain Energy	Green Mountain Energy Electricity	1.3¢/kWh	—	50% small hydro, 50% wind	Green-e
NY	Rochester Gas & Electric/Community Energy	Catch the Wind/New-Wind Energy	2.5¢/kWh	—	100-kWh blocks of new wind	—
NY	Suburban Energy Services/ Sterling Planet	Sterling Green Renewable Electricity	1.5¢/kWh	—	40% new wind, 30% small hydro, 30% bioenergy	—
PA	Energy Cooperative of Pennsylvania (10)	EcoChoice 100	2.78¢/kWh	—	89% landfill gas, 10% wind, 1% solar	Green-e
PA	Energy Cooperative of Pennsylvania (10)	Wind Energy	2.5¢/kWh	—	wind	—
PA	PECO Energy/Community Energy (10)	PECO Wind	2.54¢/kWh	—	100-kWh blocks of new wind	—
PA	PEPCO Energy Services (10)	Green Electricity 10%, 51% or 100% of usage	3.7¢/kWh (for 100% usage)	—	100% renewable	—
PA	PEPCO Energy Services (10)	NewWind Energy 51% or 100% of usage	4.48¢/kWh (for 100% usage)	—	100% new wind	—

State	Company	Product Name	Residential Price Premium[1]	Fee	Resource Mix[2]	Certification
RI	Narragansett Electric/ Community Energy, Inc.	NewWind Energy 50% or 100% of usage	2.0¢/kWh	—	50% small hydro, 50% new wind	Green-e
RI	Narragansett Electric/ People's Power & Light	New England Green-Start RI 50% or 100% of usage	1.5¢/kWh	—	69% small hydro, 30% new wind, 1% new solar	Green-e
RI	Narragansett Electric/ Sterling Planet	Sterling Supreme 100%	1.98¢/kWh	—	40% small hydro, 25% biomass, 25% new solar, 10% wind	Environmental Resources Trust
TX	Gexa Energy (11)	Gexa Green	-1.1¢/kWh	—	100% renewable	—
TX	Green Mountain Energy Company (11)	100% Wind Power: Reliable Rate or Month-to-Month	1.46¢/kWh	$5.34/mo.	wind	—
TX	Green Mountain Energy Company (11)	Pollution Free: Reliable Rate or Month-to-Month	-0.03¢/kWh	$5.34/mo.	wind and hydro	—
TX	Reliant Energy (11)	Renewable Plan	-1.1¢/kWh	—	wind	—
VA	PEPCO Energy Services (12)	Green Electricity 10%, 51% or 100% of usage	4.53¢/kWh (for 100% usage)	—	landfill gas	—
VA	PEPCO Energy Services (12)	NewWind Energy 51% or 100% of usage	5.33¢/kWh (for 100% usage)	—	new wind	—
VA	Washington Gas Energy Services/Community Energy	New Wind Energy Certificates	2.5¢/kWh	—	100 kWh blocks of new wind	—

Notes

1. Prices updated as of July 2005 and may also apply to small commercial customers. Prices may differ for large commercial/ industrial customers and may vary by service territory.

2. New is defined as operating or repowered after January 1, 1999, based on the Green-e TRC certification standards.

3. Offered in PEPCO service territory. Product prices are for renewal customers based on annual average costs for customers in PEPCO's service territory (6.8¢/kWh).

4. Price premium is for Central Maine Power service territory based on standard offer of 7.13¢/kWh.

5. Product offered in Baltimore Gas and Electric and PEPCO service territories. Price is for PEPCO service territory based on price to compare of 6.55¢/kWh.

6. Price premium is based on a comparison to the Cape Light Compact's standard electricity product.

7. Green Mountain Energy offers products in Conectiv, JCPL, and PSE&G service territories. Product prices are for PSE&G (price to compare of 6.503¢/kWh).

8. Price premium is based on a comparison to ConEdison Solutions' standard electricity product in the ConEdison service territory.

9. Price premium is for Niagara Mohawk service territory. Program only available in Niagara Mohawk service territory. Premium varies depending on energy taxes and usage.

10. Product prices are for PECO service territory (price to compare of 6.21¢/kWh).

11. Product prices are based on price to beat of 12.1¢/kWh for TXU service territory (specifically Dallas, Texas, except where noted). Except for Gexa Green, which is listed in price per kWh, prices based on 1000 kwh of usage monthly, and include monthly fees.

12. Products are available in Dominion Virginia Power service territory.

Source: This table has been reprinted from National Renewable Energy Laboratory Technical Report NREL/TP-38994, Green Power Marketing in the United States: A Status Report, 8th ed. (October 2005) by Lori Bird and Blair Swezey.

SAMPLE REQUEST FOR QUALIFICATIONS (RFQ) FOR ON-SITE SOLAR PHOTOVOLTAIC SYSTEM INSTALLERS

Adapted from: Vote Solar (www.votesolar.org)

Section 1: Background and Objectives

A. Background

Why College is pursuing qualified turnkey installers for PV projects

B. Objectives

The objective of this Request for Qualifications (RFQ) is to identify and select the most qualified turnkey photovoltaic system supplier to begin negotiating for the design, qualification, assembly, test, shipping and installation, startup and monitoring of up to __ kW of fully functioning grid-connected PV systems over a two- to three-year period.

Section 2: Solicitation Process

Each respondent to the RFQ must demonstrate that it satisfies the minimum requirements described in Section 3 in order to be selected as an eligible Respondent. The response must meet the requirements in Section 4 and must adequately address all questions in Section 5.

Responses to this RFQ must be submitted in writing, signed by an authorized officer or an agent of the respondent. College must receive ___ hard copies and one electronic copy of the respondent's package no later than the close of business day on [date]. Responses submitted after this date cannot be accepted, and responses that are incomplete or do not conform to the requirements of this RFQ will not be considered.

College intends to select one respondent from the qualified RFQ bidders list to implement its turnkey PV system installation plans over a two- to three-year period. Responses shall be submitted to:
[contact information]

All questions related to this RFQ shall be directed in writing no later than [date[to: [contact information]

College's Code of Ethics provides:

- During the process leading to the award of any contract, no member of staff of the College shall knowingly communicate any matter relating to the contract or selection process with any party financially interested in the contract, or any officer or employee of that party, unless the communication is part of the process expressly described in the request for qualifications or other solicitation invitation. Any applicant or bidder who knowingly participates in a communication that is prohibited by this section may be disqualified from the contract award.

- The process leading to the award of contract means the period between release of a request for qualifications or bids through award of the contract.

- The procedures and prohibitions prescribed by this code of ethics shall not apply to communications that are incidental, exclusively social, or do not involve the College or its business.

Key action dates are as follows:
    Request for Qualifications published: [date]
    Bidders conference and written questions submitted: [date]
    Answers publicly published by: [date]
    RFQ proposals due: [date]
    RFQ review by College: [date]
    Contract negotiation phase: [date]
    Master contract(s) awarded: [date]

Section 3: Assumptions and Minimum Project Requirements that Must Be Included in Your Proposal

Proposals submitted in response to this RFQ must be as specific as possible concerning each of the areas identified herein, including obligations of each party as envisioned by the respondent. Each respondent must provide sufficient information to enable College to understand the overall proposal, the service(s) to be provided, and the potential adverse impacts of the proposal. The College reserves the right to deem any proposal as nonresponsive and to give it no further consideration. The College also reserves the right to request clarification and or additional information from any respondent.

While specific sites have not yet been identified, for the purposes of responding to the pricing section of this RFQ, respondents are asked to make the following general project assumptions:

- Turn-key grid-connected PV projects
- [state]-based installations in the [local area] area
- Assume one or both of the following two project types and fill out information related to each in Appendix A:
  - A 250 kW flat, roof-mounted installations on facilities with less than 10 floors
  - A 1 MW ground-mounted tracking installation on a covered reservoir

- Assume 'clean' installation (with no shading and no roof protrusions if rooftop system)
- Easy access during working hours
- 480 volt electrical panel located within 100 feet of the installed system
- Prevailing wage rates for all installation labor

Both parties agree to negotiate in good faith to adjust actual pricing as necessary for specific sites to accommodate unique site requirements. Price changes will be implemented via change-order based on mutual consent of both parties.

All products and components outlined herein must conform to the following codes, standards, and rating methodologies:

- PV system must be compliant with the requirements of the state energy agency's renewable energy incentive program.
- In particular, PV modules specified in the RFQ must be certified by the state energy commission's certification program.
- Rated PV system capacity—must be specified in direct current (DC) kilowatts peak STC and PTC:
  - The STC rating, or standard test conditions rating, assumes direct current, standard test conditions (kWdc-stc). It is also referred to as kilowatts peak, or kWp. Specific PV module manufacturer maximum and minimum power data must be specified for this rating.
  - The PTC rating, or PV USA Test Conditions rating, is based on 1000 Watts/square meter solar irradiance, 20 degree Celsius ambient temperature, and 1 meter/second wind speed.
- The mathematical method for specifying PV system output in kWh must be specified for each of the following steps:
  - Calculate effective alternating current (AC) power of the proposed module type, from the rated kWdc-stc. AC losses due to wiring, soiling, and power conditioning unit (PCU) (inverter) efficiency must be taken into consideration. For full list of AC losses to consider please see Appendix B.
  - Calculate the annual kWh of the system using the PVFORM algorithm, an algorithm developed by Sandia National Laboratories in the early 1990s. Please use the same AC loss assumptions in Appendix B.
  - Specify annual degradation expected over 20 years.
- UL certification.
- National Electrical Code—NFPA 2002.
- Must comply with wind uplift requirements per the American Society of Civil Engineers Standard for Minimum Design Loads for Buildings and Other Structures (ASCE 7), and must be able to withstand design wind speeds of at least 100 mph (3-second gusts).
- All outdoor enclosures should be at minimum rated NEMA 3R.
- Occupational Health and Safety Administration (OSHA) directives.
  - For roof-mounted installations:
    PV array adds no more than 8 pounds per square foot to the facility roof structure in the array area.
  - Rooftop system components should achieve Uniform Building Code (UBC) fire code rating of B or better.

Warranty and Service Contract Requirements

1) All respondents must offer comprehensive on-site training in PV system operations, safety and maintenance consistent with warranty and service contract provisions.

2) The Respondent's standard warranty coverage should be at least 5 years for systems and 20 years for PV modules and provide:

 a) Annual on-site system inspection, including:
 - system testing (including a check of the operating current of each electrical string)
 - adjustment and routine maintenance

 b) System performance monitoring and historical data access for customer via secure Web site. Data should include: system energy and power production, ambient temperature, wind speed, and insolation

 c) Daily system monitoring by Respondent, including:
 - reporting of problems to customer
 - dispatch of resources for expeditious resolution of problems

Section 4: Comprehensive List of RFQ Proposal Components for All Bidders

1. Transmittal Letter

Please write a transmittal letter signed by a party authorized to sign binding agreements for projects of the nature ultimately contemplated by this RFQ. The letter shall clearly indicate that the respondent has carefully read all the provisions in the RFQ.

2. Project Team Qualification

Please provide the following information:
 a) Identify the team leader for the entire proposal, and his/her full contact information
 b) Identify each entity, person or firm involved in the proposal and their role e.g., design, installation, permitting, equipment supply by component, operations, and maintenance
 c) Identify the lead person responsible for each of the entities or firms described in 1b.
 d) Provide both an organizational chart and a description of responsibilities for each person or firm, and an overall project organization chart

3. Respondent Qualifications— Provide a complete profile of your firm, including

 a) Year founded
 b) Status (private/ publicly-held)
 c) Number of employees
 d) States and countries in which you do business
 e) Target customers (residential, commercial, industrial, government etc.)
 f) Organizational structure
 g) Resumes or bios of personnel to be directly involved with the development of the proposed systems
 h) Audited financial statements for the most recent three years
 i) Letter from bonding company on respondent's performance to date

4. Respondent Experience and References

a)   Overview of your company's grid connected PV experience

(i) MWp of grid connected PV installed to date

Total

By application (roof mounted, vs. ground mounted)

(ii) Average grid-connected installed PV system size during the last three years

b)   Experience: List of 5 or more grid-connected PV projects installed over the last three years that exceed 100 kWp. Include for each:

(i) Exact role(s) your organization performed for the project (eg. lead contractor, electrical subcontractor, design, consulting etc)

(ii) Location

(iii) Application description (roof vs. ground mounted)

(iv) Product name/type

(v) Customer name

(vi) Date installed, including length of time from bid acceptance to project completion

(vii) PV module used

(viii) kWdc-STC and kWac-PTC rating

(ix) Cumulative kWh produced since system installation

(x) Current status of system (operational yes/no)

c)   For the systems described above, please provide five active U.S.-based large grid-connected (>100 kWdc) customer references and their contact information

d)   Has your firm or any of the executive officers of your firm been a party to a lawsuit involving the performance of any equipment it has installed? If so, please include a summary of the issues and the status of the lawsuit.

5. Product/technology description

Respondents shall:

a)   State that their systems will comply with all of the requirements of Section 3, or list the items that would not comply and state why.

b)   Describe the technology (or technologies) that your company typically proposes for rooftop and/or ground-mounted applications including at least the following information:

c)   Photovoltaic module description, brand(s) and model numbers

(i) Inverter type and brand(s) and efficiency (in percent)

(ii) Structural materials

(iii) Balance of system components

(iv) Installed weight per square foot

c)   For each technology described above, please provide evidence that your technology and equipment is commercially proven as evidence by completed projects.

d)   For each technology described above, please describe any other benefits your system provides that other system might not provide, but only if such benefits can be readily measured and confirmed by an independent engineering study.

e)   For each technology described above, please provide information about any potentially adverse effects. For example, for rooftop systems, does your system typically penetrate the roof? If so, please describe, including expected number of penetrations per square foot, and plans to mitigate their effect.

f)   Indicate the typical degradation experienced in the field, and, if empirical evidence is not available, project the degradation rates for the useful life of the panels (but not in excess of 30 years).

6. Pricing

Using Appendix A as a template, for each PV system product/application, please provide turnkey system pricing information in $/kW (kWdc-stc) and ($/kWdc-ptc) for a 250 kWp roof-mounted PV system and/or a 1 MWp ground-mounted tracking system.

   a) Do not include any sales tax or performance/payment bond fees.
   b) Provide the estimated total kWh output of the system over 20 years using the methodology suggested in Section 4. Clearly list the AC loss assumptions in Appendix B.
   c) Clarify any pricing assumptions inherent in your bid at the time of submittal, and describe any market forces that could occur in the next 6 months to 1-year time frame that could affect those assumptions.
   d) Assume a Notice to Proceed will be signed on [date.]

Please provide an overview of your proposed system output performance verification methodology. Is it web-based? What does the end-user interface look like? Please also indicate whether or not you can comply with the minimum system performance and monitoring requirements set forth in Section 3, under Warranty and Service Requirements.

Typical project schedule and timing—For evaluation purposes, please submit a schedule for a typical 250 kW rooftop project indicating the expected milestones, with each task referenced from the notice to proceed.

Section 5: Evaluation Criteria

Principal evaluation criteria include the following:

Prior experience in developing, designing and constructing turnkey grid-connected PV projects in a timely manner (minimum of five 100kWp functioning grid-connected PV projects)—25 percent

Price in $/kWh (based on 20 years of output) and $/kWdc-stc—20 percent

Overall quality of the response to the RFQ—10 percent

Building-friendly or structure-friendly product/system design (no or few penetrations, and weighs < 8lbs/sf)—10 percent

Safe, sturdy product/system design (ASCE 7 compliant, UBC fire code rating of B or better)—10 percent

Ability to provide required warranty obligations—10 percent

For roof-mounted systems: Incremental passive savings to the facility (e.g., energy conservation delivered to the building)—5 percent

Ability to provide user-friendly, web-based performance monitoring services—5 percent

Aesthetics of system— 5 percent

The College reserves the right, at its sole discretion, to accept a response that does not satisfy all requirements but which, in the College's sole judgment, sufficiently demonstrates the ability to produce, delivery, design, permit and install a substantial volume of turnkey grid-connected PV projects and to satisfy the major requirements set forth in this RFQ. The College reserves the right to interview any or all respondents to this RFQ, or to ask for additional information or clarifications. The College expects to complete its evaluation process to select qualified contractors, but reserves the right to change key dates and action as the need arises.

Section 6: General Rules

1. No obligation—This RFQ does not obligate the College to establish eligibility for any respondents, or to issue any subsequent RFPs or to enter into any agreements. The College reserves the right to cancel or re-issue this RFQ at any time, and to solicit qualifications through any other appropriate method.

2. Rejection of Proposals—The College may reject any response that it deems to be incomplete, unresponsive, significantly inaccurate in its representation or which is unacceptable to the College in the context of this RFQ.

3. One proposal per organization—a company or nonprofit may submit only one response to this RFQ. However, a respondent may be a subcontractor to any number of other respondents that may submit responses to this RFQ.

4. Substitutions—Respondents may substitute or alter their responses subsequent to the submission date only if such changes are approved in writing by the College.

5. Cost of Proposal and Non Compensation—Each respondent is solely responsible for all costs associated with responding to this RFQ. The College will not in any event reimburse any respondent for any costs associated with this RFQ.

6. Delivery of Proposals—Each respondent is solely responsible for assuring a timely submittal of its response. Late responses will not be accepted.

7. Withdrawal of Proposal—Reponses to this RFQ maybe withdrawn after submission by written request of the respondent's authorized representative prior to the date and time specified for response submissions.

8. Disposition of Proposals, Confidential Information—All submittals and the information therein become the property of the College upon submittal. Proposals shall be returned only at the College's sole discretion and at the Respondent's expense. The College will employ reasonable efforts to hold portions of the responses specifically identified and marked as confidential in confidence to the extent permitted by law, for a period of one year beyond the required date of submittal. Upon request, the College will return or destroy all confidential information that is a trade secret or privately held company's financial information.

Appendix A

Pricing Table

Module type(s):_____

Type of technology: Crystalline___ Thin film___

Power output warranty term: 10 years__ 20 years__ 25 years__ Other____

Expected annual degradation in output: ____ percent

PV System Product/Technology Description _____

Maximum power conditioning unit (PCU) (inverter) efficiency _____ percent

PCU Brand _____

Description of other benefits provided by your product/application _____

Pricing Table (assume a purchase of 5 MW of PV installations over the next two to three years; if more than one PV module is to be used, fill out table for each module):

    System Type in kWp (kW dc-stc)

    Expected total cumulative kWh output over 20 years

    Price in $/Cumulative kWh Price in $/kWp (kW dc-stc)

    One 250 kW roof-mount

    One 1 MW ground-mount tracking

Appendix B

AC Loss Assumptions—Backup for AC Power Rating and kWh Output Calculation

Loss factors are used to convert theoretical DC output at STC to actual AC output at PTC. Please indicate the AC loss assumptions your kWh output calculation. Please be sure to include efficiency factors for the following average annual loss factors:

    DC Cabling

    Connections

    Module Coefficient of Temperature Calculation

    Module Mismatch

    Power Conditioning Unit (inverter)

    Soiling

    Shading Losses

    Tracking Losses

    Transformer Losses

    AC Wiring

    Auxiliary Loads

SAMPLE WIND POWER PURCHASE AGREEMENT

This Power Purchase Agreement ("Agreement") is made and entered into this _____ day of _____, 200_, by _____ ("Utility"), and _____("College") each hereinafter sometimes referred to individually as "Party" or both referred to collectively as the "Parties."

In consideration of the mutual covenants set forth herein, the Parties agree as follows:

1.   Scope and Purpose of Agreement

This Agreement describes the terms and conditions under which the Utility will purchase electric power from the wind turbine generator ("WTG") described in the attached Interconnection Agreement. The technical terms and definitions used in this agreement are defined in Exhibit A. The following exhibits are specifically incorporated into and made a part of this Agreement:

Exhibit A: Technical Definitions

Exhibit B: Contract Payments

2.   Description of College's Wind Turbine Generator and Interconnection

The College and Utility are Parties to an Interconnection Agreement that prescribes the design and operation of an interconnection of the College's WTG to the Utility's electric distribution system. A description of the WTG, its associated equipment, and the interconnection facilities that connect to the Utility is attached to the Interconnection Agreement as Exhibit B and by reference is made a part of this Agreement.

The College has received "Qualifying Facility" status for the WTG from the Federal Energy Regulatory Commission. The WTG is expected to start Commercial Operation in (season or approximate date).

3.   Obligations And Deliveries

3.1.   College's and Utility's Obligations. College will sell and deliver and Utility will purchase and accept all of the Output of the WTG delivered to the Interconnection Point, subject to the terms of this Agreement. There are no minimum or maximum output requirements. Utility acknowledges and understands that wind is an intermittent resource and that the output of the project, which is dependent on wind and other factors outside College's control, will constantly vary. College makes no representation or guarantee regarding the particular amount of energy or time of delivery; provided however, College agrees to sell all of WTG Output to Utility and, except as otherwise permitted under this Agreement, shall not sell any of the Output to a third party.

3.2. Payment for Output. Subject to the terms of this Agreement, Utility will pay College for the Output of the WTG at the (contract price) set forth in Exhibit B. Exhibit B may be revised from time to time as necessary to reflect significant changes in the Utility's Avoided Cost Rate.

3.3. Renewable/Environmental Attribute. There (is/is not) a presumption that any Renewable/Environmental Attribute is included as part of the WTG Output purchased under this Agreement, unless and to the extent that such Renewable/Environmental Attribute is specifically identified in Exhibit B.

3.4. Operations and Maintenance. College will be responsible for operating and maintaining the WTG, including College interconnection facilities up to the Point of Interconnection, in compliance with the terms of the Interconnection Agreement and Good Utility Practice. College will comply with all applicable local, state, and federal laws, regulations, and ordinances presently in effect or that may be enacted during the term of the Agreement.

4. Billing and Payment

4.1. Billing Period. Unless otherwise specifically agreed upon by the Parties, the calendar month will be the standard period for all billing under this Agreement. As soon as practicable after the end of each month, College will read the meter and send Utility an invoice for the payment obligations incurred under this Agreement during the previous month. The invoice shall contain the current and previous meter readings and the meter shall be accessible for reading by both Parties.

4.2. Timeliness of Payment. Unless otherwise agreed by the Parties, all invoices shall be due and payable in accordance with College's invoice instructions on or before the later of the twentieth (20th) day of each month, or twentieth (20th) day after receipt of the invoice or, if such day is not a Business Day, then on the next Business Day. Utility will make payments to the account designated by College. Any amounts not paid by the due date will be deemed delinquent and will accrue interest at the Interest Rate, such interest to be calculated from and including the due date to, but excluding the date the delinquent amount is paid in full.

4.3. Disputes and Adjustments of Invoices. The following procedures shall apply to all billing disputes. Utility may, in good faith, dispute the correctness of any invoice or any adjustment to an invoice rendered under this Agreement, or adjust any invoice for any arithmetic or computational error within six (6) months of the date the invoice or adjustment to an invoice was rendered. In the event an invoice or portion thereof or any other claim or adjustment arising hereunder is disputed, payment of the undisputed portion of the invoice shall be required to be made when due, with notice of the objection given to the invoicing Party. Any invoice dispute or invoice adjustment shall be in writing and shall state in reasonable detail the basis for the dispute or adjustment. Payment of the disputed amount shall not be required until the dispute is resolved. Upon resolution of the dispute, any required payment shall be made within two (2) Business Days of such resolution, along with interest accrued at the Interest Rate from and including the due date to, but excluding the date paid. Inadvertent overpayments shall be returned upon request or deducted by College from the next month's invoiced payments, with interest accrued at the Interest Rate from and including the date of such overpayment to, but excluding the date repaid or deducted by the College. Any dispute with respect to an invoice is waived unless the other Party is notified in accordance with the time limits established in this Section 4.3. If an invoice is not rendered within the designated time period, the right to payment for such performance is waived.

4.4. Payment Adjustment Following Correction of Metering Error. To the extent that the meter was inaccurate during a period for which payment has already been made to College by Utility, College will use the corrected measurements to recalculate the amount due either as an additional payment to College or a refund to Utility for the period of the inaccuracy. If College owes a refund to Utility, College will remit such payment not later than thirty (30) days after determination of the amount due, or at Utility's option, such refund may take the form of an offset to Utility's payments under the next regular monthly invoice. If Utility owes an additional payment to College, Utility will remit such payment not later than thirty (30) days after receiving notice of the amount due, either as a separate payment or together with the next regular monthly payment. Adjustments for metering errors shall be subject to the provisions of Section 4.7 of the Interconnection Agreement.

5.   **Audit**

Subject to reasonable conditions for preserving confidentiality, each Party has the right, at its sole expense and during normal working hours, to examine the records of the other Party to the extent reasonably necessary to verify the accuracy of any statement, charge or computation made pursuant to this Agreement. If requested, a Party shall provide to the other Party statements evidencing the Output delivered to the Interconnection Point. If any such examination reveals any inaccuracy in any statement, the necessary adjustments in such statement and the payments thereof will be made promptly and shall bear interest calculated at the Interest Rate from the date the overpayment or underpayment was made until paid; provided, however, that no adjustment for any statement or payment will be made unless objection to the accuracy thereof was made prior to the lapse of six (6) months from the rendition thereof, and thereafter any objection shall be deemed waived.

6.   **Limitations of Remedies, Liability and Damages**

Each Party's liability to the other Party for any loss, cost, claim, injury, liability, or expense, including court costs and reasonable attorney's fees, relating to or arising from any act or omission in its performance of this Agreement, shall be limited to the amount of direct damage or liability actually incurred. In no event shall either Party be liable to the other Party for any indirect, incidental, special, consequential, or punitive damages of any kind whatsoever.

Utility shall not be liable to College in any manner, whether in tort or contract or under any other theory, for loss or damages of any kind sustained by College resulting from existence of, operation of, or lack of operation of the WTG, or termination of the Interconnection Agreement, provided such termination is consistent with the terms of the Interconnection Agreement, except to the extent such loss or damage is caused by the negligence or willful misconduct of the Utility.

7.   **Indemnification**

College and Utility shall each indemnify, defend and hold the other, its directors, officers, employees, and agents (including, but not limited to, affiliates and contractors and their employees), harmless from and against all liabilities, damages, losses, penalties, including reasonable attorney fees, claims, demands, suits and proceedings of any nature whatsoever for personal injury (including death) or property damages to unaffiliated third parties that arise out of or are in any manner connected with the performance of this Agreement by that Party except to the extent that such injury or damages to unaffiliated third parties may be attributable to the negligence or willful misconduct of the Party seeking indemnification.

8.   **Force Majeure**

If a Force Majeure Event prevents a Party from fulfilling any obligations under this Agreement, such Party will promptly notify the other Party in writing, and will keep the other Party informed on a continuing basis of the scope and duration of the Force Majeure Event. The affected Party will specify in reasonable detail the circumstances of the Force Majeure Event, its expected duration, and the steps that the affected Party is taking to mitigate the effects of the event on its performance. The affected Party will be entitled to suspend or modify its performance of obligations under this Agreement but only to the extent that the effect of the Force Majeure Event cannot be mitigated by the use of reasonable efforts. The affected Party will use reasonable efforts to resume its performance as soon as practicable.

9.   **Governmental Charges**

College shall pay or cause to be paid all Government Charges arising prior to the Point of Interconnection. Utility shall pay or cause to be paid all Governmental Charges arising at and from the Point of Interconnection (other than ad valorem, franchise or income taxes that are related to the sale of the Output and are, therefore, the responsibility of the College). In the event College is required by law or regulation to remit or pay Governmental Charges that are Utility's responsibility hereunder, Utility will reimburse College for such Governmental Charges within thirty (30) days following written notice from College. If Utility is required by law or regulation to remit or pay Governmental Charges that are College's responsibility hereunder, Utility may deduct the amount of any such Governmental Charges from the sums due to College under Section 3 of this Agreement. Nothing shall obligate or cause a Party to pay or be liable to pay any Governmental Charges for which it is exempt under the law, and nothing shall obligate a Party to forego its right to challenge any Governmental Charges in the applicable legal forum, provided that responsibility for any expenses associated with such a legal challenge shall be the sole responsibility of the Party undertaking such a challenge.

10.   **Severability**

If any provision or portion of this Agreement shall for any reason be held or adjudged to be invalid or illegal or unenforceable by any court or any regulatory body of competent jurisdiction, such portion or provision shall be deemed separate and independent, and the remainder of this Agreement shall remain in full force and effect.

11.   **Communications**

The Parties may also designate operating representatives to conduct the daily communications that may be necessary or convenient for the administration of this Agreement. Such designations, including names, addresses, and phone numbers may be communicated or revised by one Party's Notice to the other in accordance with Section 21.

12.   **Right of Access, Equipment Installation, Removal and Inspection**

The College shall notify the Utility at least two days prior to the initial testing and operation of the WTG. The Utility may send a qualified person to the premises of the College at or immediately before the time of initial testing and operation to inspect the interconnection and observe the commissioning of the WTG. Following the initial inspection process described above, at reasonable hours, and upon reasonable notice, or at any time without notice in the event of an emergency or hazardous condition, Utility shall have access to College's premises for any reasonable purpose in connection with the performance of the obligations imposed on it by this Agreement.

13.   **Effective Term and Termination Rights**

13.1.   This Agreement becomes effective when executed by both Parties and shall continue in force for 20 years from the commercial date of operation of the WTG. The Parties may at their option agree to extend the Term by a written amendment to this Agreement signed by both Parties. Applicable provisions of this Agreement will continue to be in effect after the conclusion of the Term or upon earlier termination to the extent necessary to provide for final billings and adjustments, including payment of any money due and owing by either Party. This Agreement can be terminated under one of the following conditions:

13.1.1.   College may terminate this Agreement at any time, by giving the Utility two (2) years' written notice;

13.1.2.   Either Party may terminate by giving the other Party at least sixty (60) days' prior written notice that the other Party is in default of any of the material terms and conditions of the Agreement, so long as the notice specifies the basis for termination and there is reasonable opportunity to cure the default.

13.1.3. Either Party may terminate by giving the other Party at least sixty (60) days' prior written notice if the Interconnection Agreement is terminated.

14.   Governing Law

This Agreement was executed in the State of _____ and must in all respects be governed by, interpreted, construed, and enforced in accordance with the laws thereof. This Agreement is subject to, and the Parties' obligations hereunder include, maintaining and operating in full compliance with all valid, applicable federal, State, and local laws or ordinances, and all applicable rules, regulations, orders of, and tariffs approved by, duly constituted regulatory authorities having jurisdiction.

15.   Assignments

The College will obtain the written consent of the Utility before the College assigns this Agreement to a corporation, another entity with limited liability, or another person. Such consent will not be withheld or unreasonably delayed, provided College can demonstrate that the entity is reasonably capable of performing the obligations of this Agreement. The College will supply the Utility with all reasonable information concerning such entity that is requested by the Utility.

16.   Confidentiality

If either Party requests, the metering, inspection, and maintenance records will be considered to be confidential information and will not be provided to any third party (other than the Party's employees, lenders, counsel, accountants, or advisors who have a need to know such information and have agreed to keep such terms confidential) without the written consent of the other Party and subject to the right and obligations of any application law governing public records.

17.   Dispute Resolution

17.1.   All disputes and differences pertaining to or arising out of this Agreement or the breach thereof, which cannot be settled by mutual consent of both Parties and which are not subject to the jurisdiction of the Federal Energy Regulations Commission FERC or the State Utility Board, may be submitted to arbitration at the request of either Party. Such arbitration will be held in (state) , and, except as otherwise provided herein, shall be conducted in accordance with the provisions of the arbitration rules of the American Arbitration Association in the absence of contrary agreement by the Parties. Copies of any such request shall be served on all Parties and such request shall specify the issues in dispute and summarize the complaining Party's claim with respect thereto.

17.2.   Each Party shall designate one arbitrator. In the event that the two arbitrators so designated cannot agree upon a third arbitrator, the third arbitrator shall be selected in accordance with such rules. All arbitrators shall be experienced in utility and/or wind generation issues.

17.3.   The three-person panel designated in accordance with subsection 17.2 of this Section shall conduct a hearing and within 30 days thereafter (unless such time is extended by agreement of the Parties) shall render a decision, and shall notify the Parties in writing of the decision, stating the reasons for such decision and separately listing findings of fact and conclusion of law. The panel shall have no power to amend or otherwise revise this agreement. Subject to such limitation, the decision of the panel shall be final and binding, except that any Party may petition a court of competent jurisdiction for review of errors of law. The decision shall determine and specify how the expense of the arbitration shall be borne.

17.4.   The Parties shall continue to perform their obligations under this Agreement during the pendency of any proceeding provided for by this Section.

18.   Amendment and Notification

This Agreement can only be amended or modified by a written document signed by both Parties.

19.   Entire Agreement

This Agreement constitutes the entire Agreement between the Parties and supersedes all prior agreements or understandings, whether verbal or written. It is expressly acknowledged that the Parties may have other agreements covering other services not expressly provided for herein, which agreements are unaffected by this Agreement.

20.   Non-Waiver

None of the provisions of this Agreement shall be considered waived by a Party unless such waiver is given in writing. The failure of a Party to this agreement to insist, on any occasion, upon strict performance of any provision of this agreement will not be considered to waive the obligations, rights, or duties imposed on the Parties.

21.   No Third-Party Beneficiaries

This agreement is not intended to and does not create rights, remedies, benefits of any character whatsoever in favor of any persons, corporations, associations, or entities other than the Parties, and the obligations herein assumed are solely for the use and benefit of Parties, their successors in the interest and, where permitted, their assigns.

22.   Notices

Any written notice, demand, or request required or authorized in connection with this Agreement ("Notice") shall be deemed properly given if delivered in person or sent by first-class mail, postage prepaid, to the person specified below:

If to College: _____

If to Utility: _____

A Party may change its address for Notices at any time by providing the other Party Notice of the change in accordance with Section 22.

23.   Signatures

IN WITNESS WHEREOF, the Parties have caused this Agreement to be signed by their respective duly authorized representatives.

_____   _____

Utility Representative   College Representative

Date: _____   Date: _____

EXHIBIT A

Definitions for Terminology Used in the Agreement

- Agreement—This is the Power Purchase Agreement between the Utility and the College.

- Avoided Cost Rate—This is the payment rate for the WTG Output that reflects the Utility's avoided wholesale power purchase costs. The Avoided Cost Rate is described in Exhibit B.

- Business Day—This is any day except a Saturday, Sunday, or a Federal Reserve Bank holiday. A Business Day shall open at 8:00 a.m. and close at 5:00 p.m. local time for the relevant Party's principal place of business.

- College—This is the [name of college] in [city, state], which receives retail electric service from [name of utility].

- Commercial Operation—This is the date when the WTG is legally and operationally capable of producing energy and commencing deliveries to the Utility.

- Contract Price—This is the total price in $U.S. to be paid by the Utility to the College for the purchase of the Output.

- Force Majeure Event—For purposes of this Agreement, a "Force Majeure Event" means any event: (a) that is beyond the reasonable control of the affected Party; and (b) that the affected Party is unable to prevent or provide against by exercising reasonable diligence, including but not limited to the following events or circumstances, but only to the extent they satisfy the preceding requirements: acts of war, public disorder, insurrection, terrorism, or rebellion; floods, hurricanes, earthquakes, lightning, storms, and other natural calamities; explosions or fires; strikes, work stoppages, or labor disputes; embargoes; and sabotage.

- Good Utility Practice—Any of the practices, methods, and acts engaged in or approved by a significant portion of the electric utility industry during the relevant time period, or any of the practices, methods, and acts which, in the exercise of reasonable judgment in light of the facts known at the time the decision was made, could have been expected to accomplish the desired result at a reasonable cost consistent with good business practices, reliability, safety, and expedition. Good Utility Practice is not intended to be limited to the optimum practice, method, or act to the exclusion of all others, but rather to the acceptable practices methods, or acts generally accepted in the region.

- Governmental Charges—Taxes that are imposed by any governmental authority on or with respect to the Output that arise out of the operation of this Agreement.

- Interconnection—The physical connection of WTG to the Utility's electric distribution system in accordance with the requirements of this Agreement so that parallel operation can occur.

- Interconnection Agreement—A separate agreement that sets forth the contractual conditions under which the Utility and the College agree that WTG may be interconnected with the Utility's system.

- Interest Rate—The term Interest Rate is the same per annum rate of interest that the Utility is obligated under statute to pay its customers for any amounts due.

- Output—This is the electrical energy produced by the WTG and delivered at the Point of Interconnection, together with any related firm generating capacity as may be defined by the Utility, its wholesale power supplier, or the power pool. The Output includes all energy produced from the time of initial testing until the wind turbine is retired by the College. The Output also includes Renewable/Environmental Attributes that have been specifically identified in Exhibit B.

- QF—This is a Qualifying Facility as defined by the Federal Energy Regulatory Commission and it includes the WTG and all interconnection facilities owned and used by the College.

- Party or Parties—means either the Utility or the College or both.

- Point of Interconnection—The point where the College's interconnection equipment connects to the Utility's electric distribution system as shown on Exhibit B of the Interconnection Agreement.

- Renewable/Environmental Attribute—This means any attribute associated with the Output that is related to the WTG's renewable or environmental characteristics and that has independent value, including without limitation renewable certification for purposes of satisfying a renewable portfolio standard or other federal, state or local regulatory program designed to encourage procurement of renewable resources, "green" credits, air quality and emission avoidance credits.

- Term—means the duration of this Agreement as specified in Section 13 of the Agreement.

- WTG—The large utility scale wind turbine generator to be installed by the College.

EXHIBIT B

Contract Payments

Utility will pay College for the Output from the WTG. The payment will be calculated by multiplying the metered Output by the Avoided Cost Rate. The Avoided Cost Rate is the Utility's net savings (calculated on a per kWh basis) obtained in its wholesale power bill due to the avoidance of power purchases caused by purchasing the Output from the College's WTG. The savings in the Utility's power bill will also include any applicable demand charges, surcharges, or taxes. Initially the Avoided Cost Rate will be about ($0.035 per kWh). The Avoided Cost Rate may change from month to month to track the Utility's actual avoided costs. The Utility will provide the College with a document showing the calculation of the Utility's Avoided Cost Rate. The Utility may update the Avoided Cost Rate calculation and documentation at any time. The College will use the most recent Avoided Cost Rate provided by the Utility to calculate its invoice to the Utility.

This Avoided Cost Rate [does/does not] include the purchase of Renewable/Environmental Attributes. The College will initially retain all Renewable/Environmental Attributes from the WTG.

This Exhibit B is [the original] [or] [_____ revision] made and entered into this _____ day of _____, 200__ .

By: _____

Date: _____

SAMPLE AGREEMENT FOR INTERCONNECTION OF WIND TURBINE GENERATOR WITH UTILITY

This Interconnection Agreement ("Agreement") is made and entered into this ___day of _____, 200__, by _____ _____ ("Utility"), and _____ ("College") each hereinafter sometimes referred to individually as "Party" or both referred to collectively as the "Parties."

In consideration of the mutual covenants set forth herein, the Parties agree as follows:

1. Scope and Purpose of Agreement

This Agreement describes only the conditions under which the Utility and the College agree that the wind turbine generator ("WTG") described in Exhibit B may be interconnected to and operated in parallel with the Utility's system. Any purchase of power from the WTG will be covered under a separate agreement. The technical terms and definitions used in this agreement are defined in Exhibit A. The following exhibits are specifically incorporated into and made a part of this Agreement:

Exhibit A: Technical Definitions

Exhibit B: Technical Information on Generator and Interconnection

2. Description of College's Wind Turbine Generator and Interconnection

A description of the WTG, its associated equipment, and the interconnection facilities that connect to the Utility is attached to and made a part of this Agreement as Exhibit B.

The College has received "Qualifying Facility" status for the WTG from the Federal Energy Regulatory Commission. The WTG is expected to start Commercial Operation [start date.]

3. Ownership and Operation of WTG and Interconnection Equipment

3.1. Subject to the terms and conditions of this Agreement, the College shall engineer, design, construct, operate and maintain the WTG and its interconnection equipment in accordance with Good Utility Practice. The College shall interconnect the WTG to Utility's electrical system and shall operate its WTG in parallel with the Utility's electric system subject to approval of the interconnection by Utility, which approval shall not be unreasonably withheld or delayed. The College will pay for all interconnection equipment, including underground cables, primary metering equipment, switches, fuses, and other necessary protective and control equipment necessary to connect to the Utility's existing distribution system. The Utility shall have the right, but not the obligation, to review the design and inspect the interconnection equipment and to inform College of any problems it believes may exist and any proposals it has for correcting any such problems.

3.2. The College shall operate the WTG in a manner that does not adversely affect the power quality for other electric customers on the Utility's distribution system. Power quality aspects include the generation of harmonic frequencies, voltage flicker, or voltage levels outside normal ranges that cause problems or complaints from other electric customers. The Institute of Electrical and Electronic Engineers (IEEE) 1547 Standard entitled "IEEE Standard for Interconnecting Distributed Resources with Electric Power Systems" will be used as a guide for the proper interconnection and operation of the WTG.

3.3. College shall, at its own expense, install, maintain, and periodically test any equipment necessary for protection against abnormal conditions such as inadvertent re-energizing of a dead line, ground or phase faults, voltages and frequencies outside of acceptable limits, or as otherwise required by Good Utility Practice and IEEE 1547 Standard.

3.4. Utility shall have the right to inspect, at its own expense and upon reasonable notice to College, any interconnection equipment for the WTG whenever such inspection may be deemed by it to be necessary. With the prior consent of College (which consent shall not be unreasonably withheld or delayed), or in cases of emergency, without prior consent by College, Utility's inspection may include testing of electric disconnect devices. If any such equipment is found not to be working properly, Utility shall promptly notify College, and College shall promptly cause it to be corrected at the College's expense. Until such correction is accomplished, Utility has the right to suspend acceptance of electricity generated by the WTG and may open the interconnection between its electric system and the WTG if, in Utility's reasonable opinion, continuing to accept such electricity would adversely affect the safe and reliable operation of its Utility's electric system. Utility's inspection of College's equipment shall not be construed as endorsing the design thereof nor as any warranty of the safety, durability or reliability of said equipment, nor shall it relieve College from its responsibility to maintain and test such equipment.

3.5. College shall be responsible for protecting the WTG and its interconnection equipment from possible damage by reason of electrical disturbances or faults caused by the operation, faulty operation, or nonoperation of Utility's electrical system, as well as of other electric systems interconnected to Utility's electrical system, and unless due to wanton, willful or grossly negligent conduct of Utility, Utility shall not be liable for any such damages so caused.

4.   Metering Equipment and Procedures

4.1. College shall purchase and install metering equipment approved by Utility, which approval shall not be unreasonably withheld, capable of time-differentiated (by hour) measurement of the electric output and consumption of the WTG in kWh and of reactive power flow with a communications port suitable for a modem. The Utility may install a more sophisticated meter for purposes of measuring the power quality and waveforms; however, the Utility will pay the additional incremental cost of these features. Metering equipment (including meter, current transformers, and voltage transformers) shall be of revenue metering accuracy. Upon installation, all metering equipment will become and will remain property of the Utility throughout the Term. Upon termination of this Agreement, College will have the option to purchase the metering equipment from the Utility for $1. The metering shall be kept under seal by the Utility, such seal to be broken only when the meters are to be tested, adjusted, modified, or relocated. In the event the College breaks a seal, College shall notify Utility as soon as practicable.

4.2. College shall make available by telephone to Utility access to meters to retrieve any metered data. College shall be responsible for all costs associated with maintaining this telephone access.

4.3.   Meters purchased by College shall be capable of separately determining on an hourly basis the backup power used when the net electric output of the WTG is negative (i.e., when the WTG's electric usage exceeds its electric generation).

4.4.   Since the WTG will be located in the service territory of Utility, the College will purchase all backup power for the WTG from under a separate agreement with Utility.

4.5.   Utility shall own, control, maintain and repair meters, conduct meter accuracy and tolerance tests and prepare all calibration reports required for equipment that measures energy transfers between College and Utility in accordance with State of _____ requirements. Utility shall be reimbursed by College for all reasonably incurred costs associated with work it performs on meters and related equipment. The College, at its own expense, may have an independent third party check the accuracy of the meter.

4.6.   Utility shall, where practicable, provide at least five (5) days notice to College prior to any test it may perform on the metering equipment. College shall have the right to have a representative present during each such test.

4.7.   If any metering equipment is found to be inaccurate by more than two percent (2 percent), such meter will be promptly corrected and the meter readings for the period of inaccuracy shall be adjusted as far as can be reasonably ascertained by using any and all relevant data and information. If the Parties cannot agree on the actual period during which the inaccurate measurements were made, the period during which the measurements are to be adjusted will be the shorter of (a) the last one-half of the period from the last previous test of the metering device to the test that found the metering device to be defective or inaccurate, or (b) the 180-day period immediately preceding the test that found the metering device to be defective or inaccurate.

5.   Responsibilities of Utility and College

Each Party will, at its own cost and expense, operate, maintain, repair, and inspect, and shall be fully responsible for, the facility or facilities which it now or hereafter may own or lease unless otherwise specified in Exhibit B. Maintenance of College's WTG and interconnection facilities shall be performed in accordance with the applicable manufacturer's recommended maintenance schedule. Utility and College shall each be responsible for the safe installation, maintenance, repair and condition of their respective lines and appurtenances on their respective sides of the Point Of Interconnection. The Utility or the College, as appropriate, shall provide interconnection facilities that adequately protect the Utility's distribution system, personnel, and other persons from damage and injury.

6.   Disconnection, Suspension, and Reduction of Power Delivery

6.1.   Upon notification from the Utility, or if recognized by College's personnel or agents, that the operation of the WTG is causing or substantially contributing to adverse power quality on the Utility's distribution system, the College shall immediately suspend or reduce electricity deliveries to the extent required to eliminate such adverse impact. An adverse power quality impact shall include the generation of harmonic frequencies, voltage fluctuations, or operating voltage levels outside normal ranges that adversely affects the safe and reliable operation of Utility's electric system. If College fails or refuses to comply with a request pursuant to this Section to suspend or reduce electric deliveries, or if giving such notice is not practicable, the Utility may trip or disconnect the WTG. If the operation of the WTG is primarily causing the condition, College shall, at its own cost, modify its electric facilities to the extent necessary to promptly resume full deliveries of electricity at a quality of electric service that does not have an adverse impact on Utility's electric system or customers. Upon College's request, the Utility will modify its electrical facilities to assist College in resuming full deliveries provided that College reimburses Utility for all costs and expenses incurred by Utility in making such modifications. If the Utility requests a suspension or reduction of deliveries pursuant to this Section, it shall, as soon as practicable, provide a written statement to the College setting forth the reasons for such suspension or reduction.

6.2.   In the event of an abnormal condition on Utility's electrical system, including but not limited to Force Majeure, emergency, safety problem, forced outage, period of routine maintenance that cannot reasonably be coordinated with College's period of maintenance or shutdown, or physical inability of Utility's system to accept electricity deliveries, College shall suspend or reduce deliveries of electricity as requested by Utility. Utility shall use reasonable efforts to remedy the abnormal condition or take other appropriate action so that full deliveries of electricity can be restored as soon as practicable. If Utility requests a suspension or reduction pursuant to this subsection, it shall, as soon as practicable, provide a written statement to College setting forth the reasons for such suspension or reduction request and the likely duration thereof.

6.3.   If any suspension or reduction of deliveries pursuant to this Section 6 is due to the gross or willful negligence of Utility or the material violation of this Agreement by Utility, Utility shall be liable to the College solely for the verifiable net revenues and incentives that College would have received for any lost sales of generation.

6.4.   Utility may temporarily disconnect the WTG without prior notice in cases where, in the sole judgment of the Utility, continued interconnection will endanger persons or property. The Utility shall have the right to temporarily disconnect the WTG to make repairs on the Utility's system. When possible, the Utility shall provide the College with reasonable notice and reconnect the WTG as soon as practicable.

7.   Prior Authorization

For the mutual protection of the College and the Interconnection Provider, the connections between the Utility's service wires and the College's service entrance conductors shall not be energized without prior authorization of the Utility, which authorization shall not be unreasonably withheld or delayed. The fuse disconnects will have provisions for locking with a Utility lock and will always be operated by the Utility. In no event will the College close the fuse disconnects near the Point of Interconnection. It shall be the responsibility of the College to provide the Utility with a spare set of fuses.

8.   Warranty Is Neither Expressed Nor Implied

Neither by inspection, if any, or nonrejection, nor in any other way, does the Utility give any warranty, express or implied, as to the adequacy, safety, or other characteristics of any structures, equipment, wires, appliances or devices owned, installed or maintained by the College or leased by the College from third parties, including without limitation the WTG and any structures, equipment, wires, appliances or devices appurtenant thereto.

9.   Limitations of Liability

Each Party's liability to the other Party for any loss, cost, claim, injury, liability, or expense, including court costs and reasonable attorney's fees, relating to or arising from any act or omission in its performance of this Agreement, shall be limited to the amount of direct damage or liability actually incurred. In no event shall either Party be liable to the other Party for any indirect, incidental, special, consequential, or punitive damages of any kind whatsoever.

Utility shall not be liable to College in any manner, whether in tort or contract or under any other theory, for loss or damages of any kind sustained by College resulting from existence of, operation of, or lack of operation of the WTG, or termination of the Interconnection Agreement, provided such termination is consistent with the terms of the Interconnection Agreement, except to the extent such loss or damage is caused by the negligence or willful misconduct of the Utility.

10.   Indemnification

College and Utility shall each indemnify, defend and hold the other, its directors, officers, employees and agents (including, but not limited to, affiliates and contractors and their employees), harmless from and against all liabilities, damages, losses, penalties, claims, demands, suits and proceedings of any nature whatsoever for personal injury (including death) or property damages to unaffiliated third parties that arise out of or are in any manner connected with the performance of this Agreement by that Party except to the extent that such injury or damages to unaffiliated third parties may be attributable to the negligence or willful misconduct of the Party seeking indemnification.

11. Force Majeure

If a Force Majeure Event prevents a Party from fulfilling any obligations under this Agreement, such Party will promptly notify the other Party in writing, and will keep the other Party informed on a continuing basis of the scope and duration of the Force Majeure Event. The affected Party will specify in reasonable detail the circumstances of the Force Majeure Event, its expected duration, and the steps that the affected Party is taking to mitigate the effects of the event on its performance. The affected Party will be entitled to suspend or modify its performance of obligations under this Agreement but only to the extent that the effect of the Force Majeure Event cannot be mitigated by the use of reasonable efforts. The affected Party will use reasonable efforts to resume its performance as soon as practicable.

12. Insurance

Prior to the start of the installation and continuing through the duration of this Agreement, the College shall purchase and maintain insurance that will protect the College from liability and claims for injuries and damages, which may arise out of or result from the College's operations under the Agreement. The insurance coverage will include: a) commercial general liability with a minimum limit of $2,000,000 for each occurrence, b) comprehensive automobile liability coverage, including all owned, nonowned, and hired vehicles, with a minimum limit of $1,000,000, and c) workers compensation and employers liability insurance at statutory limits. Utility shall have the right to inspect or obtain a copy of the original policies of insurance for each such coverage

13. Severability

If any provision or portion of this Agreement shall for any reason be held or adjudged to be invalid or illegal or unenforceable by any court or regulatory body of competent jurisdiction, such portion or provision shall be deemed separate and independent, and the remainder of this Agreement shall remain in full force and effect.

14. Communications

The Parties may also designate operating representatives to conduct the daily communications that may be necessary or convenient for the administration of this Agreement. Such designations, including names, addresses, and phone numbers may be communicated or revised by one Party's Notice to the other in accordance with Section 24.

15. Right of Access, Equipment Installation, Removal and Inspection

College shall provide Utility with periodic updates of the construction status of the WTG and the expected completion of construction, the expected energization and testing, and the expected Commercial Operation date. Initially, the time between such updates shall not exceed 30 calendar days. During the 30 days prior to energization, the time between such updates shall not exceed 7 calendar days. The College shall notify the Utility at least two days prior to the initial testing and operation of the WTG. The Utility may send a qualified person to the premises of the College at or immediately before the time of initial testing and operation to inspect the interconnection and observe the commissioning of the WTG. Following the initial inspection process described above, at reasonable hours, and upon reasonable notice, or at any time without notice in the event of an emergency or hazardous condition, Utility shall have access to College's premises for any reasonable purpose in connection with the performance of the obligations imposed on it by this Agreement.

16. Effective Term and Termination Rights

16.1. This Agreement becomes effective when executed by both Parties and shall continue in force for 20 years from the Commercial Operation of the WTG, or until terminated under one of the following conditions:

16.1.1.   College may terminate this Agreement at any time, by giving the Utility two (2) years' written notice;

16.1.2. Utility may terminate upon failure by the College to generate energy from the WTG in parallel with the Interconnection Provider's system by the later of two years from the date of this agreement or twelve months after completion of the interconnection; or

16.1.3. Either Party may terminate by giving the other Party at least sixty (60) days' prior written notice that the other Party is in default of any of the material terms and conditions of the Agreement, so long as the notice specifies the basis for termination and there is reasonable opportunity to cure the default.

16.2. Upon termination of this Agreement, the WTG will be disconnected from the Utility's electric system.

The termination of this Agreement shall not relieve either Party of its liabilities and obligations, owed or continuing at the time of the termination.

17. Governing Law

This Agreement was executed in the State of _____ and must in all respects be governed by, interpreted, construed, and enforced in accordance with the laws thereof. This Agreement is subject to, and the Parties' obligations hereunder include, maintaining and operating in full compliance with all valid, applicable federal, State, and local laws or ordinances, and all applicable rules, regulations, orders of, and tariffs approved by, duly constituted regulatory authorities having jurisdiction.

18. Assignments

The College will obtain the written consent of the Utility before the College assigns this Agreement to a corporation, another entity with limited liability, or another person. Such consent will not be withheld or unreasonably delayed, provided College can demonstrate that the entity is reasonably capable of performing the obligations of this Agreement. The College will supply the Utility with all reasonable information concerning such entity that is requested by the Utility.

19. Confidentiality

If either Party requests, the metering, inspection, and maintenance records will be considered to be confidential information and will not be provided to any third party (other than the Party's employees, lenders, counsel, accountants or advisors who have a need to know such information and have agreed to keep such terms confidential) without the written consent of the other Party and subject to the rights and obligations of any applicable law governing public records.

20. Dispute Resolution

20.1. All disputes and differences pertaining to or arising out of this Agreement or the breach thereof, which cannot be settled by mutual consent of both Parties and which are not subject to the jurisdiction of the Federal Energy Regulations Commission (FERC) or the [State] Utility Board, may be submitted to arbitration at the request of either Party. Such arbitration will be held in the State of _____, and, except as otherwise provided herein, shall be conducted in accordance with the provisions of the arbitration rules of the American Arbitration Association in the absence of contrary agreement by the Parties. Copies of any such request shall be served on all Parties and such request shall specify the issues in dispute and summarize the complaining Party's claim with respect thereto.

20.2. Each Party shall designate one arbitrator. In the event that the two arbitrators so designated cannot agree upon a third arbitrator, the third arbitrator shall be selected in accordance with such rules. All arbitrators shall be experienced in utility and/or wind generation issues.

20.3. The three-person panel designated in accordance with subsection 19.2 of this Section shall conduct a hearing and within 30 days thereafter (unless such time is extended by agreement of the Parties) shall render a decision, and shall notify the Parties in writing of the decision, stating the reasons for such decision and separately listing findings of fact and conclusion of law. The panel shall have no power to amend or otherwise revise this agreement. Subject to such limitation, the decision of the panel shall be final and binding, except that any Party may petition a court of competent jurisdiction for review of errors of law. The decision shall determine and specify how the expense of the arbitration shall be borne.

20.4. The Parties shall continue to perform their obligations under this Agreement during the pendency of any proceeding provided for by this Section.

21. Amendment and Notification

This Agreement can only be amended or modified by a written document signed by both Parties.

22. Entire Agreement

This Agreement constitutes the entire Agreement between the Parties and supersedes all prior agreements or understandings, whether verbal or written. It is expressly acknowledged that the Parties may have other agreements covering other services not expressly provided for herein, which agreements are unaffected by this Agreement.

23. Non-Waiver

None of the provisions of this Agreement shall be considered waived by a Party unless such waiver is given in writing. The failure of a Party to this agreement to insist, on any occasion, upon strict performance of any provision of this agreement will not be considered to waive the obligations, rights, or duties imposed on the Parties.

24. No Third Party Beneficiaries

This agreement is not intended to and does not create rights, remedies, benefits of any character whatsoever in favor of any persons, corporations, associations, or entities other than the Parties, and the obligations herein assumed are solely for the use and benefit of Parties, their successors in the interest and, where permitted, their assigns.

25. Notices

Any written notice, demand, or request required or authorized in connection with this Agreement ("Notice") shall be deemed properly given if delivered in person or sent by first class mail, postage prepaid, to the person specified below:

If to College:

Attention:

Address:

Phone:

If to Utility:

Attention:

Address:

Phone:

A Party may change its address for Notices at any time by providing the other Party Notice of the change in accordance with Section 25.

26. Signatures

IN WITNESS WHEREOF, the Parties have caused this Agreement to be signed by their respective duly authorized representatives.

_____   _____

Representative of Utility   Representative of College

Date: _____   Date: _____

EXHIBIT A
Definitions for Terminology Used in the Agreement

- *Agreement*—means this Interconnection and Parallel Operation Agreement between the Utility and the College.

- *College*—[name of College], which receives retail electric service from [name of Utility Company].

- *Commercial Operation*—This is the date when the WTG is legally and operationally capable of producing energy and commencing deliveries to the Utility.

- *Force Majeure Event*—For purposes of this Agreement, a "Force Majeure Event" means any event: (a) that is beyond the reasonable control of the affected Party; and (b) that the affected Party is unable to prevent or provide against by exercising reasonable diligence, including but not limited to the following events or circumstances, but only to the extent they satisfy the preceding requirements: acts of war, public disorder, insurrection, terrorism or rebellion; floods, hurricanes, earthquakes, lightning, storms, and other natural calamities; explosions or fires; strikes, work stoppages, or labor disputes; embargoes; and sabotage.

- *Good Utility Practice*—Any of the practices, methods, and acts engaged in or approved by a significant portion of the electric utility industry during the relevant time period, or any of the practices, methods, and acts which, in the exercise of reasonable judgment in light of the facts known at the time the decision was made, could have been expected to accomplish the desired result at a reasonable cost consistent with good business practices, reliability, safety, and expedition. Good Utility Practice is not intended to be limited to the optimum practice, method, or act to the exclusion of all others, but rather to the acceptable practices, methods, or acts generally accepted in the region.

- *Indemnification*—means protection against or being kept free from loss or damage.

- *Interconnection*—The physical connection of WTG to the Utility's electric distribution system in accordance with the requirements of this Agreement so that parallel operation can occur.

- *Interconnection Agreement ("Agreement")*—The Agreement sets forth the contractual conditions under which the Utility and the College agree that WTG may be interconnected with the Utility's system.

- *Party* or *Parties*—means either the Utility or the College or both.

- *Point of Interconnection*—The point where the College's interconnection equipment connects to the Utility's electric distribution system as shown on Exhibit B.

- *QF*—This is a Qualifying Facility as defined by the Federal Energy Regulatory Commission and it includes the WTG and all interconnection facilities owned and used by the College.

- *Term*—means the duration of this Agreement as specified in Section 16 of the Agreement.

- *Utility*—[name of Utility Company], which operates an electric distribution system, and its agents or permitted successors and assigns.

- *WTG*—means the wind turbine generator installed by the College as described herein.

EXHIBIT B
Technical Information on Generator and Interconnection

Type of Generator: Vestas NM82 1650 kW wind turbine

Generator Voltage: 600 volts

Expected Average Annual Generation: 5.9 million kWh

Wind Turbine Step Up Transformer: 2000 kVA, 600 Volt/12.47 kV

Location of Wind Turbine: SW¼ of the SW¼ of Section 13

The College will interconnect the WTG to the Utility's 12.47 kV electric distribution system near the intersection of 7th Avenue South and South 19th Street. This will be the Interconnection Point. A set of fused disconnects and a primary metering package will be installed at the Interconnection Point. Underground 12.47 kV 3-phase cables will be buried on the west side of South 19th Street southward to the wind turbine site. Attachment 1 shows a one-line electrical drawing of the interconnection to the distribution system.

Wind Utility Consulting has completed a Power Quality Impact Analysis of the wind turbine on the Utility's distribution system. That report is included as Attachment 2. The analysis indicates that the WTG is expected to have no adverse impact on the power quality of the Utility's system.

INFORMATION RESOURCES

Federal Government

U.S. Department of Energy (DOE)
National Renewable Energy Laboratory, www.nrel.gov/
Office of Energy Efficiency and Renewable Energy, www.eere.energy.gov/
University Buildings, www.eere.energy.gov/buildings/info/university/index.html
Rebuild America, college and university sectors, www.rebuild.gov/sectors/col_uni/index.asp

U.S. Environmental Protection Agency (EPA)
Green Power Partnership, www.epa.gov/greenpower/index.htm
Colleges and universities sector, www.epa.gov/sectors/colleges/
U.S. DOE/EPA Energy Star for Higher Education, www.energystar.gov/index.cfm?c=higher_ed.bus_highereducation

Higher Education Sustainability Organizations

Associated Colleges of the South (ACS), Environmental Initiative, www.colleges.org/~enviro
Association for the Advancement of Sustainability in Higher Education, www.aashe.org
Campus Consortium for Environmental Excellence, www.c2e2.org
Council of Environmental Deans & Directors, http://ncseonline.org/CEDD/
Green Campus Consortium of Maine, www.megreencampus.com
Higher Education Climate Action Project, www.heclimateaction.org
Indiana Consortium for Education towards Sustainability, www.ices.ws
Michigan Higher Education Partnership for Sustainability, www.ecofoot.msu.edu
New Jersey Higher Education Partnership for Sustainability, www.njheps.org, www.njheps.org/toolkit.pdf
Northeast Campus Sustainability Coalition
Pennsylvania Consortium for Interdisciplinary Environmental Policy, www.paconsortium.state.pa.us
Second Nature, www.secondnature.org
South Carolina Sustainable Universities Initiative, www.sc.edu/sustainableu

United Negro College Fund, Building Environmental Stewardship, www.uncfsp.org/bes
University Leaders for a Sustainable Future, www.ulsf.org
Upper Midwest Association for Campus Sustainability, www.umacs.org

Nonprofit Organizations

Alliance to Save Energy, Green Campus Program, www.ase.org/section/program/greencampus
American Council for an Energy-Efficient Economy, www.aceee.org
American Council on Renewable Energy, www.acore.org
APPA, www.appa.org
Association of University Leaders for a Sustainable Future, www.ulsf.org
Center for Resource Solutions, www.resource-solutions.org
Clean Air-Cool Planet, www.cleanair-coolplanet.org/for_campuses.php
Energy Services Coalition, www.energyservicescoalition.org/
Environmental Resources Trust, Inc., www.ert.net
International City County Management Association, www.icma.org
National Association of College and University Business Officers, www.nacubo.org
National Association of Energy Service Companies, www.naesco.org
National Wildlife Federation, Campus Ecology Program, www.nwf.org/campusecology
Pew Center on Global Climate Change, www.pewclimate.org
Renewable Energy Policy Project, www.crest.org
Society for College and University Planning, www.scup.org
World Resources Institute, Green Power Market Development Group, www.thegreenpowergroup.org

Industry Associations

American Wind Energy Association, www.awea.org
Biomass Energy Research Association, www.bera1.org
Geothermal Energy Association, www.geo-energy.org
National Hydrogen Association, www.hydrogenus.com
Solar Energy Industries Association, www.seia.org
U.S. Fuel Cell Council, www.usfcc.com

State Clean Energy Funds

Clean Energy States Alliance, www.cleanenergystates.org
California: California Energy Commission, www.energy.ca.gov
Connecticut: Connecticut Clean Energy Fund, www.ctcleanenergy.com
Illinois: Illinois Clean Energy Community Foundation, www.illinoiscleanenergy.org
Massachusetts: Massachusetts Renewable Energy Trust, www.mtpc.org/RenewableEnergy/
Minnesota: Xcel Energy Renewable Development Fund, www.xcelenergy.com
New Jersey: New Jersey Clean Energy Program, www.njcep.com
New York: Long Island Power Authority Clean Energy Initiative, www.lipower.org; New York State
    Energy Research and Development Authority, www.nyserda.org
Ohio: Energy Loan Fund, www.odod.state.oh.us/cdd/oee/energy_loan_fund.htm
Oregon: Energy Trust of Oregon, www.energytrust.org

Pennsylvania: MetEd Sustainable Energy Fund of the Berks County Community Foundation, www.bccf.
org/pages/gr.energy.html; Penelec Sustainable Energy Fund of the Community Foundation for the
Alleghenies, www.cfalleghenies.org; Sustainable Development Fund, www.trfund.com/sdf;
Sustainable Energy Fund of Central Eastern Pennsylvania, www.sustainableenergyfund.org;
West Penn Power Sustainable Energy Fund, Inc., www.wppsep.org
Rhode Island: Rhode Island Renewable Energy Fund, www.riseo.state.ri.us/riref.html
Wisconsin: Focus on Energy, www.focusonenergy.com

Financial Information and Resources

Chicago Climate Exchange, www.chicagoclimatex.com
Database of State Incentives for Renewable Energy (DSIRE), www.dsireusa.org
U.S. Department of Energy, Office of Energy Efficiency and Renewable Energy
   Federal grants and funds for renewable energy and energy efficiency:
   www.eere.energy.gov/cleancities/pdfs/epa_funding_opps_04.pdf
   Guidebook for states on buying power from wind generators as part of State Implementation Plans
   (SIP): www.eere.energy.gov/windandhydro/windpoweringamerica/pdfs/wpa/sips_model.pdf
Local Government Commission
   Sample RFPs for solar installations, www.lgc.org/spire/rfps.html.
National Renewable Energy Lab
   RETFinance, http://analysis.nrel.gov/retfinance/login.asp (calculates cost of energy of renewable
   electricity generation technologies such as biomass, geothermal, solar, and wind)
Renewable Energy Finance & Investment Network
   Directory of financing sources, www.acore.org/pdfs/REFIN_2004_lores.pdf

On-Site Renewable Energy Resources

California Energy Commission, *California Interconnection Guidebook,* www.energy.ca.gov/distgen/in-
terconnection/guide_book.html. Connecting to the utility grid in California, but useful for colleges and
universities in other states.

Center for Resource Solutions, *Regulator's Handbook on Tradable Renewable Certificates.* http://crs2.net/
handbook/TRC_Handbook.htm. For states interested in selling their RECs, but also useful for colleges
and universities.

Interstate Renewable Energy Council and North Carolina Solar Center, *Connecting to the Grid,* www.
irecusa.org/pdf/guide.pdf.

USDA Forest Service Forest Products Laboratory, "Primer on Wood Biomass for Energy," May 2004, ww.fpl.
fs.fed.us/tmu/pdf/primer_on_wood_biomass_for_energy.pdf; "Wood Biomass for Energy," April 2004, www.
fpl.fs.fed.us/documnts/techline/wood_biomass_for_energy.pdf

Calculators

DOE/EPA Energy Star

- Cash Flow Opportunity (CFO) calculator, www.energystar.gov/ia/business/cfo_calculator.xls
- Portfolio Manager, www.energystar.gov/index.cfm?c=evaluate_performance.bus_portfoliomanager

DOE, National Renewable Energy Laboratory

www.nrel.gov/analysis/power_databook/calculators.html.
Includes:

- Renewable Energy Conversion
- Renewable Energy Growth Estimator
- Photovoltaic Land Area Estimator
- Wind Power Land Area Estimator

DOE, Rebuild America, www.rebuild.org/lawson/Calculators.asp
Includes:

- Fuel Conversion Calculator
- Energy Intensity Calculator
- Internal Rate of Return Calculator
- Life Cycle Cost Calculator
- Net Present Value Calculator
- Opportunity Calculator

ABOUT THE SPONSORS

The New Jersey Higher Education Partnership for Sustainability (NJHEPS)
c/o NJIT—York CEES
138 Warren Street
Newark, NJ 07102
973.642.4881
Fax 973.642.7170
www.njheps.org
John L. Cusack
Executive Director
jcusack@njheps.org

The New Jersey Higher Education Partnership for Sustainability (NJHEPS) is a consortium of 40+ New Jersey colleges and universities with the goal of encouraging and facilitating the use and practice of sustainability concepts in higher education and society. NJHEPS concentrates on energy-efficient operations, integrating sustainability into curriculum, and improving the sustainability of local communities.

U.S. Department of Energy's Office of Energy Efficiency and Renewable Energy (EERE)
www.eere.energy.gov
The Department of Energy's Office of Energy Efficiency and Renewable Energy (EERE) leads the charge for far-reaching technological change, to conserve and diversify the energy sources used to fuel America, and to lay the foundation for independence from imported oil. EERE pursues a diverse portfolio of research, development, and demonstration, all with one ultimate aim: to bring you a prosperous future where energy is abundant, reliable, and affordable.

International City/County Management Association
777 North Capitol St. NW, Suite 500
Washington D.C. 20002
202.289.4262
Fax 202.962.3500
www.icma.org
Barbara Yuhas
202.962.3539
byuhas@icma.org

ICMA is the professional and educational organization for chief appointed managers, administrators, and assistants in cities, towns, counties, and regional entities throughout the world. Since 1914, ICMA has provided technical and management assistance, training, and information resources to its members and the local government community. The management decisions made by ICMA's nearly 8,000 members affect more than 100 million individuals in thousands of communities--from small towns with populations of a few hundred to metropolitan areas serving several million.

Chevron Energy Solutions
12980 Foster Drive, Suite 400
Overland Park, KS 66213
913.563.3500
Fax 913.563.3565
www.chevronenergy.com
Steven Spurgeon, Manager of Marketing Communications
913.563.3609
sspurgeon@chevron.com

Chevron Energy Solutions' mission is to help institutions improve facility infrastructure, use less energy and ensure reliable, high quality power critical to keeping indoor environments comfortable, productive and safe. Our "total energy services" approach includes everything from engineering analysis and design to construction management, commissioning and monitoring, and verification of energy savings. Chevron has expertise in traditional energy saving applications, cutting edge power applications such as cogeneration, hydrogen fuel cells and distributed generation, as well as experience in renewable power sources such as photovoltaics, wind, and landfill gas.

American Council on Renewable Energy (ACORE)
1629 K Street, NW
Washington, DC 20006
202.393.0001
Fax: 202.393.0606
www.acore.org
Tom Weirich, Business Development Associate
202.393.0001 x7582
weirich@acore.org

American Council on Renewable Energy (ACORE), a 501(c)(3) membership nonprofit organization headquartered in Washington, D.C., is dedicated to bringing renewable energy into the mainstream of the U.S. economy and lifestyle through convening, information, and communications programs. ACORE provides a common platform for the wide range of interests in the renewable energy community including industries, associations, utilities, universities, end users, professional service firms, financial institutions, and government agencies. ACORE serves as a forum through which the parties work together on common interests. Membership information is available at www.acore.org.

National Wildlife Federation, Campus Ecology Program
11100 Wildlife Center Drive
Reston, VA 20190-5362
800.332.4949
Fax: 703.438.6468
www.nwf.org/campusecology
Julian Keniry, Director, Campus and Community Leadership
703.438.6322
keniry@nwf.org

NWF's Campus Ecology Program helps teams of staff, students, and faculty improve the forecast for people and wildlife by designing greener campuses. With Energy Action, a coalition of more than a dozen organizations, Campus Ecology has launched the Campus Climate Challenge to support and recognize campuses that demonstrate how the U.S. can reduce climate altering emissions by at least 30 percent below 2005 levels by the year 2020 (or 2 percent per year on average) and better prepare students for leadership and career opportunities. To realize this vision, Campus Ecology offers a variety of resources and support, including NWF's Campus Ecology Recognition, yearbooks, and other nationally recognized publications, fellowships, teleconferences and more.